AF361921

Simulation and Analysis of High Voltage Engineering in Power Systems

Simulation and Analysis of High Voltage Engineering in Power Systems

Editor

Dan Doru Micu

MDPI • Basel • Beijing • Wuhan • Barcelona • Belgrade • Manchester • Tokyo • Cluj • Tianjin

Editor
Dan Doru Micu
Technical University of Cluj-Napoca
Romania

Editorial Office
MDPI
St. Alban-Anlage 66
4052 Basel, Switzerland

This is a reprint of articles from the Special Issue published online in the open access journal *Energies* (ISSN 1996-1073) (available at: https://www.mdpi.com/journal/energies/special_issues/S_A_HVE_PS).

For citation purposes, cite each article independently as indicated on the article page online and as indicated below:

LastName, A.A.; LastName, B.B.; LastName, C.C. Article Title. *Journal Name* **Year**, *Volume Number*, Page Range.

ISBN 978-3-0365-4365-9 (Hbk)
ISBN 978-3-0365-4366-6 (PDF)

Contents

About the Editor

Dan Doru Micu

Dan Doru Micu has been with the Technical University of Cluj-Napoca, an "Advanced Research and Education University", since 1999 where he is currently a Full Professor, the Director of the Energy Transition Research Center, and a PhD Supervisor. Between 2006 and 2010 he was a Scientific Expert at the Managerial Agency of Scientific Research at Romanian Ministry of Research and Innovation, and starting from 2020, he has been selected as a Scientific Officer at the National Council for Scientific Research in the same Ministry. He has lectured at more than 60 Universities all over the world. In 2017 to 2018, he was a Fulbright Senior Fellow at the University of Florida, US. Since 2019, he has also worked as a Part-time Professor at Beijing Jiaotong University, China. He has authored and coauthored more than 250 scientific papers published in peer-reviewed journals and presented at international conferences. He has been the project manager of 15 research grants obtained by international competitions (8 Horizon2020 Projects) in the field of power systems, numerical energy analytics tools, and energy efficiency in buildings and industry.

Article

Optimized Protection of Pole-Mounted Distribution Transformers against Direct Lightning Strikes

Mahdi Pourakbari-Kasmaei [1],*, Farhan Mahmood [2] and Matti Lehtonen [1]

[1] Department of Electrical Engineering and Automation, Aalto University, Maarintie 8, 02150 Espoo, Finland; matti.lehtonen@aalto.fi

[2] Department of Electrical Engineering, University of Engineering and Technology, Lahore 39161, Pakistan; fmahmood@uet.edu.pk

* Correspondence: Mahdi.Pourakbari@aalto.fi; Tel.: +358-46-538-48-72

Received: 27 July 2020; Accepted: 22 August 2020; Published: 24 August 2020

Abstract: Direct lightning strikes on overhead phase conductors result in high overvoltage stress on the medium voltage (MV) terminals of pole-mounted transformers, which may cause considerable damage. Therefore, introducing an efficient protection strategy would be a remedy for alleviating such undesirable damages. This paper investigates the optimized protection of MV transformers against direct lightning strikes on the phase conductors. To this end, first, the impacts of grounding densities (number of grounded intermediate poles between every two successive transformer poles) on the probability of overvoltage stress on transformer terminals are investigated. Then, the implications of guy wire, as a supporting device for ungrounded intermediate poles, on reducing the overvoltage stress on transformers, are studied. Finally, the role of a surge arrester in mitigating the overvoltage stress of non-surge-arrester-protected transformer poles is scrutinized. The investigations are conducted on a sample MV network with 82 wood poles comprising 17 pole-mounted transformers protected by spark gaps. To provide in-depth analysis, two different poles, namely creosote- and arsenic-impregnated poles, are considered under wet and dry weather conditions. A sensitivity analysis is performed on grounding distances and on a combination of guy wire and grounded intermediate poles while taking into account soil ionization. The results provide a clear picture for the system operator in deciding how many grounded intermediate poles might be required for a system to reach the desired probabilities of transformers experiencing overvoltage stress and how the surge arrester and guy wires contribute to mitigating undesirable overvoltage stress.

Keywords: direct lightning; guy wire; grounding density; pole-mounted transformer; spark gap; surge arrester

---

1. Introduction

A medium voltage (MV) distribution network with overhead lines may experience high overvoltage due to direct lightning strikes. Such transient overvoltages are the primary source of damage to MV equipment, especially power transformers. They cause stress on the insulation system that gradually results in failure or severe damage [1]. The results of a survey on transformer reliability by Cigre Study Committee A2.37 [2] show that the largest numbers of failures are caused by natural disasters. In this report, since all the transformers were protected by surge arresters, the failure rate due to lightning was around 3%. Furthermore, the highest failure rate can be attributed to design, manufacturing and aging, with 32%, while 29% of causes remained unknown. This high unknown failure rate and even the failure rate corresponding to the aging category might be the result of the accumulation of prior small damages due to transient overvoltages [1,3].

The protection level of MV transformers in a network depends on several factors, such as insulation level [4,5], protection devices (e.g., spark gap, surge arrester, etc.) [6,7], and grounding conditions [8], among others. More often than not, MV transformers are protected by spark gaps against lightning overvoltage [9]. These devices are cheap; however, when triggered, they cause an interruption in the electricity supply to the consumers [10]. On the other hand, a surge arrester is considered the most appropriate device to provide adequate protection for power transformers against direct and indirect lightning phenomena [11]. Nevertheless, as concluded by experimental analyses in [12], the protective performance of surge arresters depends on the steepness of the lightning surge as well as the earthing system. In some works, the spark gaps were considered as the only protection device for MV transformers, while there exist various works that considered different combinations of the spark gap and surge arrester. In [13], first, the authors investigated the impacts of the spark gap and the surge arrester separately; then, a combination of these devices was considered on the MV side of the transformer. This work deduced that an efficient way to reduce the overvoltage might be combining the spark gap and a surge arrester with a lower rating (which is cheaper) in series to reach the desired result. In [14], another combination of spark gap and surge arrester was considered such that the spark gap was installed on the MV side of the transformer, while on the low voltage (LV) side, a surge arrester was considered. The outcome of this work was mitigating the transmitted lightning overvoltage from the MV side to the LV side and toward the low voltage electricity consumers. Unlike [14], the authors in [15] investigated the effects of installing surge arresters on the MV side of the transformer while installing a spark gap on the LV side. It was shown that installing the spark gap on the LV side limited the overvoltage stress on the MV side. Practical experiments show that one of the most efficient protection strategies for transformers against lightning overvoltages is installing surge arresters on both the medium voltage (MV) and low voltage (LV) sides [16]; however, this might be an expensive solution for MV networks. Therefore, to install limited numbers of surge arresters in optimal locations, the placement of the surge arrester has been studied in some works. Heuristic algorithms such as the genetic algorithm [6], fuzzy logic [17], etc., are the most commonly used methods to find the optimal placement for a surge arrester, while some works have proposed innovative approaches to the allocation problem. The authors in [18] proposed a novel methodology, namely direct discharge crossing (DDC), to analyze the critical condition of the network and then install surge arresters in optimal locations to prevent equipment failure. The methodology offered a lightning performance function based on two main factors, namely the amplitude of overvoltage and the number of flashovers due to direct lightning, and then recommended the locations of surge arresters to mitigate the lightning overvoltage stress. It was shown in [19] that even installing a surge arrester may not always prevent flashovers on the transformers, and this may cause minor damage in the internal circuit of the transformer, and, over time, it may cause complete failure of insulation. This is most probably due to the inherent behavior of surge arresters that minimizes the effects of the steep wavefront of lightning overvoltages, while they are not capable of eliminating the surges generated from reaching transformers [20].

Considering an appropriate model for the grounding system plays a crucial role in conducting high voltage studies and helps in obtaining more practical results. More often than not, 22 kV medium voltage networks are not supported by overhead ground wires against lightning strikes. A direct lightning strike to the phase conductor of ungrounded poles results in a multi-phase fault [21,22]. Therefore, in MV networks, for the sake of providing better protection for the transformers against external overvoltage surges, as well as due to safety requirements, the transformer poles are grounded. Direct lightning imposes a very high current that, in grounded poles, goes toward the ground and may cause soil ionization as a result of a breakdown inside the air voids existing among the soil particles [23]. Soil ionization results in decreasing the earthing resistance [24], i.e., under the ionization condition, the resistance is varied by the current. However, in most of the existing works on distribution networks, constant resistances were used to model the grounding resistance [8,25,26]. In [25], the importance of earthing resistance in the performance of surge-protecting devices installed in the LV terminal of the pole-mounted distribution transformer was investigated. In this work, different constant ground resistances were considered in order to investigate the effect of grounding resistance size on phase-to-earth overvoltage amplitude under different lightning peak currents. In [8], by considering different constant grounding resistances of surge arresters and measuring the imposed voltage and current on the LV terminals, it was concluded that the lower the resistance, the better the protective functionality of the surge arrester becomes. Experimental results also emphasize the importance of low grounding resistance for better performance of protective devices such as surge arresters [12]. However, the authors in [26], by considering different constant grounding resistances for the surge arrester installed on the MV side of the transformer in a network, concluded that, for their particular configuration, (1) increasing the grounding resistance slightly changes the protective performance of surge arresters, and (2) equipping the system with proper protective devices is much more important than decreasing the grounding resistance. Defining a suitable distance between two grounded poles is yet another critical issue. In [27], it was demonstrated that the effectiveness of a shielding wire on the mitigation of lightning-induced overvoltages depends on the value of earthing resistance, while, in [7], the authors stated that it also depends on the distance between two consecutive grounding points. Still, there is a gap in the study of the impact of grounding density between two successive grounded transformer poles on mitigating the overvoltage stress at the terminals of transformers.

According to the aforementioned literature review, and to the best of our knowledge, (a) no work so far has considered the impact of different grounding densities between two successive grounded intermediated poles (poles between two transformer poles) on the probability of different overvoltage stress levels of transformers occurring in a network, (b) the impact of guy wire on protecting the MV network against direct lightning has always been neglected, and (c) there is a gap in investigating the effects of surge arrester(s) on the overvoltage stress of adjacent non-surge-arrester-protected transformer poles as well as on the probability of overvoltage stress at the terminals of transformers.

Therefore, the main contributions of this work are as follows:

- Investigating the impact of different grounding spans (spans between two successive grounded intermediate poles) on the probability of overvoltage stress.
- Investigating the impact of guy wire, installed at the ungrounded poles, on the probability of overvoltage stress at the adjacent pole-mounted transformers.
- Modeling the transient performance of the grounding system by using a nonlinear resistance for soil ionization due to large lightning current.
- Investigating the impact of surge arrester(s), installed on the MV side of the transformers on mitigating the overvoltage stress of the adjacent non-surge-arrester-protected pole-mounted transformer.

The remainder of this paper is organized as follows. Section 2 presents the simulation setup for different components of an MV distribution network. Primary parts of the sample system are validated in Section 3. Detailed information on the case studies and the primary assumptions are provided in Section 4. Section 5 provides simulation results and in-depth analyses. Section 6 presents the concluding remarks and prospects of future works.

2. Simulation Setup

This section provides detailed information on the simulation setup related to a part of a typical 3-phase medium-voltage (MV) network. The components to be modeled in Electromagnetic Transients Program-Restructured Version (EMTP-RV) [28] are direct lightning impulse, wood poles, metallic cross arm, insulators, overhead lines, earthing system, guy wire, spark gap, and surge arrester.

2.1. Direct Lightning

In order to model the behavior of direct lightning in a software environment, a current source is used. In this paper, a Cigre current source is utilized to simulate the behavior of direct lightning [29]. The required parameters to produce the direct lightning impulse are presented in Table 1.

Table 1. Parameters of direct lightning [13].

Parameter	Symbol	Value
Maximum current (kA)	I_{max}	27.7
Front time (µs)	t_f	3.83
Maximum steepness $\left(\frac{kA}{µs}\right)$	S_m	23.3
Time to half value (µs)	t_h	77.5

Figure 1 depicts the direct lightning impulse generated by using the Cigre current source available in the EMTP-RV and by applying the values presented in Table 1.

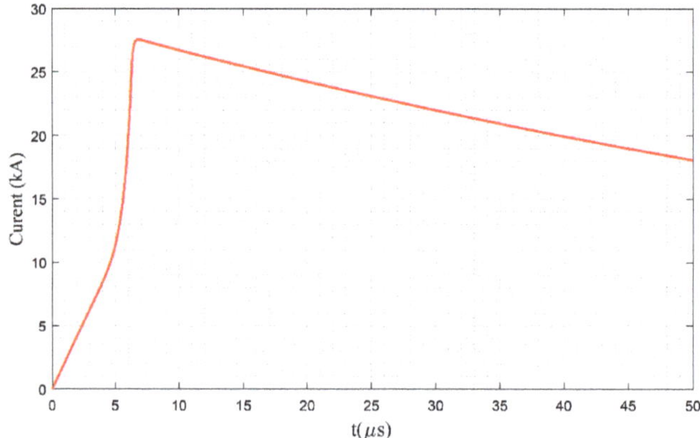

Figure 1. Lightning current impulse.

2.2. Wood Poles and Cross Arm

Figure 2 presents the dimensions and different parts of an MV wood pole. The wood poles for 3-phase overhead distribution lines—see Figure 2a—are modeled by a parallel resistor-capacitor circuit, while the metallic cross arms—see Figure 2b—are modeled by inductors [13]. The capacitance for this 7 m pole is 5.36 pF [30], and the value per unit length of inductances (to model the cross arms) is 1 µH/m. Cross arms of grounded poles are connected to the footing resistance via a lead wire (Figure 2e), while, for ungrounded poles, the lead wire is removed. The resistance of the pole highly depends on the weather conditions [31]. In this work, poles with two different impregnation materials, namely creosote and arsenic, are studied under wet and dry weather conditions. The resistances of creosote- and arsenic-impregnated poles under wet conditions in November, are 580 kΩ and 28 kΩ, respectively, while, under dry conditions in August, both poles are considered to have similar resistances of 1400 kΩ [31]. Note that these resistances also depend on the geographical characteristics and the above values have been used due to their applicability in Finland.

The occurrence of a flashover down the wood pole is modeled via a voltage-controlled switch. When the voltage over the pole exceeds the critical flashover voltage (CFO) of the pole, the switch is closed; otherwise, the switch remains open. Under the wet condition, the CFOs for creosote- and arsenic-impregnated wood poles are assumed to be 320 kV/m and 160 kV/m, respectively, while, under the dry condition, the CFO for both poles is assumed to be 400 kV/m. The arcing resistance for poles is set to be 80 Ω. In high voltage studies, the CFO of two or more components in series is the summation of the CFOs of each component [21]. Therefore, in the case of grounded poles, the CFO is simply the CFO of insulators—see Section 2.3—while, in the case of the ungrounded pole, the flashover voltage depends on the combined dielectric strength of the insulator and pole connected in series, which is the summation of the CFOs. It is noteworthy to mention that, if an ungrounded pole is supported by guy wire, the total CFO is the CFO of the insulator plus the CFO of the pole corresponding to the length from cross arm to the guy wire connection point.

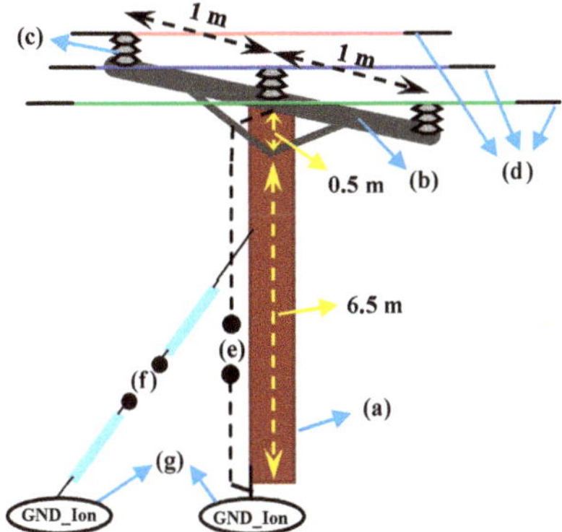

Figure 2. Dimensions and different parts of the medium voltage wood pole for a 3-phase distribution network: (**a**) pole, (**b**) cross arm, (**c**) insulator, (**d**) overhead lines, (**e**) lead wire (used for grounding), (**f**) guy wire, and (**g**) earthing system.

2.3. Modeling the Insulator

The insulators, which are located between the conductors (overhead lines) and cross arm—see Figure 2c—are modeled by considering a parallel resistor-capacitor circuit, where the values of capacitance and resistance are set to 80 pF and 25 MΩ, respectively [22].

On the other hand, it is necessary to simulate the flashover phenomena, which may occur over the insulators by lightning strikes.

In this work, the length of the insulator is 18 cm and a time-dependent model is used to model the behavior of the insulator closer to the practical situation. To do so, the flashover voltage is approximated by (1).

$$V_{flashover} = a \cdot l + \frac{b \cdot l}{(t)^{0.75}} \tag{1}$$

where $V_{flashover}$ is the flashover voltage in kV, a and b are arbitrary parameters that can be adjusted via laboratory test and, in this work, these parameters are set to 400 and 700, respectively; l is the insulator length in m, and t is the elapsed time after lightning strike in µs [22,32,33].

2.4. Overhead Line

In order to simulate the behavior of the overhead lines for the sample 3-phase MV network under direct lightning conditions, J. Marti's frequency dependent (FD) line model is used [34]. This model is available in the library of the EMTP-RV, while the technical data need to be prepared for the overhead lines. The DC resistance and outside diameter of the conductors are set to 0.509 Ω/km and 0.6 cm, respectively. The length of each span is 80 m, while, for the end points of the network, long overhead lines of 15 km are considered.

2.5. Earthing System

In order to have a more practical model, the ionization of the soil needs to be taken into account—see Figure 2g. Footing resistance is nonlinearly proportional to the value of current, such that, for low currents, there will be a constant resistance, while the resistance of a ground electrode may decrease due to ionization of the soil. In this paper, the footing resistance under high impulse current is calculated by (2) [24].

$$R = \frac{R_0}{\sqrt{1 + \frac{I_R}{I_g}}} \tag{2}$$

where R is the footing resistance at high impulse current, R_0 is the footing resistance at low current, I_R is the current through the footing resistance, and I_g is the current required to initiate the soil ionization gradient, E_0. In this work, for grounded poles, R_0 is assumed to be 17 Ω [35,36]. The limiting current I_g is calculated by (3).

$$I_g = \frac{1}{2\pi} \frac{\rho E_0}{R_0^2} \tag{3}$$

where ρ is the soil resistivity, and E_0 is the soil breakdown gradient. In this paper, it is assumed that the soil resistivity and soil breakdown gradient are 75 (Ωm) [37] and 400 kV/m [29,37], respectively.

It is noteworthy to mention that, in EMTP-RV, an existing current-controlled nonlinear resistance is used to simulate the behavior of the earthing system [28].

2.6. Guy Wire

A guy wire is a tensioned cable designed to support the poles by enhancing their stability. More often than not, guy wires are used in the dead-end, transformer poles and also in the angles of the line route. Guy wire is modeled by an equivalent constant parameter (CD) line model with a surge impedance of 450 Ω [38,39]. Guy wire earthing system can be modeled as a constant resistance since the current conducted through this wire is relatively low and does not initiate the soil ionization [40]. However, in this work, due to considering the arsenic-impregnated pole under the wet condition, which has a relatively low resistance, the earthing system described in Section 2.5 is used. The distance of the guy wire from the metallic part of the cross arm is 2 m. It is worth mentioning that, for the guy wire earthing system, the footing resistance is considered to be 40 times higher than the pole grounding resistance [38].

2.7. Spark Gap

A spark gap, which is used to protect the transformers by limiting the overvoltage amplitude, is modeled via the disruptive effect (DE) method. When the amplitude of the voltage over the spark gap exceeds the critical flashover voltage between the terminals, it triggers [41]. In some works, a simple voltage-controlled switch is used to model the behavior of the spark gap [13,14]; however, in this paper, the DE method in (4), which presents a more precise result, is used.

$$\int_{t_0}^{t} \left[\left| v_{gap}(t) \right| - v_0 \right]^k dt \geq D \tag{4}$$

where the k, V_0, and D are an empirical constant, the onset voltage of primary ionization (kV), and the disruptive effect constant (kV μs), respectively. In this paper, $V_0 = 112$ kV, $k = 0.97$, and $D = 0.01$ are used to simulate the behavior of an 8 cm sphere-sphere spark gap.

2.8. Surge Arrester

In this paper, the frequency dependent behavior of the metal oxide surge arrester is modeled according to the guidelines provided by the IEEE Working Group 3.4.11 [42]. To do so, a resistor (R), inductor (L), and capacitor (C) circuit, namely RLC circuit, presented in Figure 3, is used, in which two nonlinear resistance Ao and A1 are separated by an R-L filter. The required information to simulate a surge arrester is obtained from [43] and the voltage-current (V-I) characteristics of the nonlinear resistances are derived from [42].

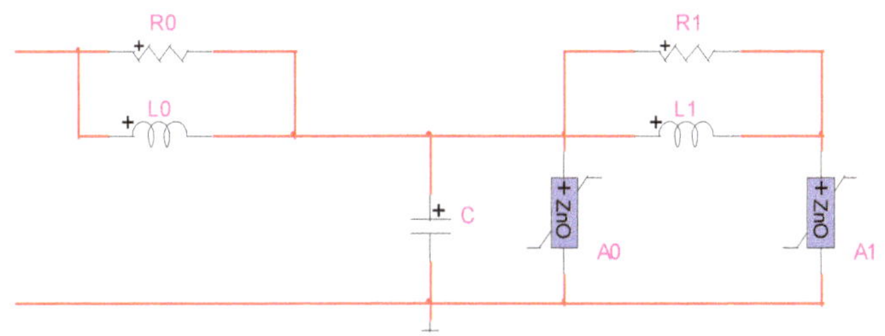

Figure 3. Frequency-dependent metal oxide surge arrester [42].

2.9. Transformer

In this work, a delta–star ($\Delta - Y$) 22/0.4 kV transformer is modeled in the EMTP-RV environment, according to Cigre Guidelines [44]. Figure 4 shows the basic building block of a transformer for one phase in EMTP-RV, while Figure 5 presents the 3-phase transformer model in which the capacitances on the primary side, between the primary and secondary sides, and on the secondary side have been considered, which play a crucial role in high voltage transient studies.

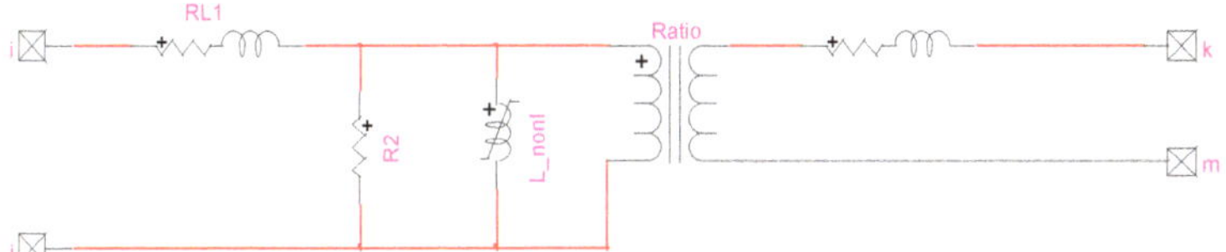

Figure 4. Basic building block for each phase of the transformer, EMTP-RV model.

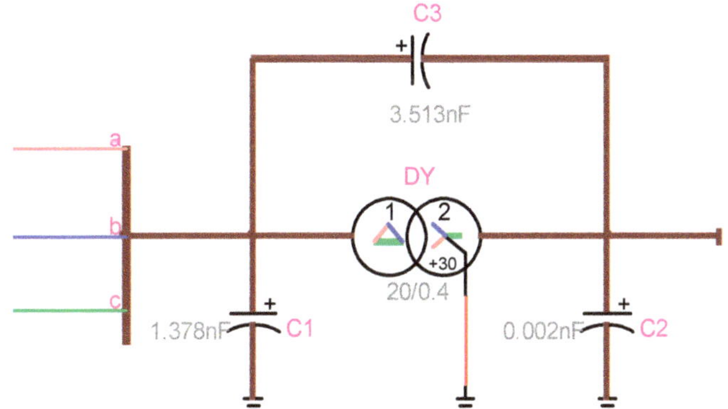

Figure 5. ΔY connection model of the transformer with measured capacitances.

2.10. Sample Network

The sample MV network in Figure 6 consists of 81 wood poles, and the length of the spans is 80 m. The length of the lines at both ends (e.g., FD line 1) is set to 15 km. The network has 17 pole-mounted transformers at every five spans (400 m). The first transformer is placed on pole #1, and the next on pole #6, and so on, while the direct lightnings are applied on every three spans, e.g., pole #1, pole #4, and so on.

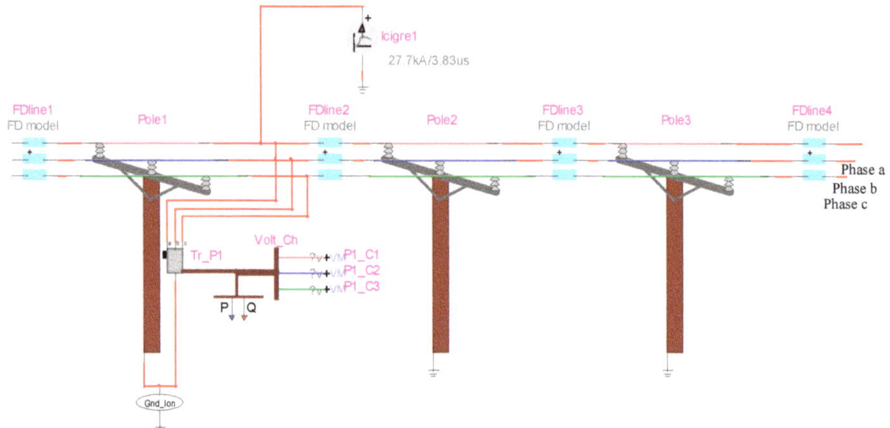

Figure 6. Sample medium voltage network.

3. Model Validation

This section provides adequate information and analyses to validate primary parts of the system, such as spark gap, transformer capacitances, insulator flashover, as well as surge arrester.

3.1. Spark Gap and Transformer

In the literature, several approaches have been proposed to obtain a good approximation of the DE parameters such as k, V_0, and D. However, the best way to adjust these parameters is to have some experimental data. In this work, an 8 cm sphere-sphere spark gap, presented in Figure 7, is used [45].

Figure 7. Adjustable sphere-sphere spark gap [45].

In order to validate the proper functionality of this spark gap, a nonstandard 125 kV overvoltage impulse is used. In Figure 8a, the green curve shows the applied impulse in a laboratory, while the blue curve stands for the applied impulse modeled in EMTP-RV.

To adjust the parameters of the DE method, more often than not, some approximation methods are used, although these methods may not always give an accurate result. As an example, in [46], these values are estimated by determining the critical flashover voltage (CFO) as $V_0 = 0.86 \times$ CFO, $D = 1.29 \times 10^{-6} \cdot$ CFO, and k = 1 that results in $V_0 = 106.1$, $D = 0.1.59$. By considering these values in our work, the spark gap was not triggered under 125 kV applied impulse and the transformer experienced the overvoltage presented in Figure 8a, while the experimental measurements show that the spark gap is triggered—see Figure 8b. Therefore, in this work, first, the V_0 is estimated via the method proposed by the Cigre Group [47] ($V_0 \approx 0.9 \cdot$ CFO ≈ 112 kV), and then a trial and error method is used to find a better agreement with the laboratory results. To this end, as suggested in [47], the k is considered to be lower than 1 and then the parameter D is adjusted. Therefore, in this paper, the parameters are set as follows: $V_0 = 112$ kV, k $= 0.97$, and $D = 0.01$.

On the other hand, the performance and triggering of the spark gap highly depends on the characteristic of the transformer, especially the capacitances on the primary side (C1), on the secondary side (C2), and between the primary and secondary sides (C3) of the transformer—see Figure 5. To obtain these values, laboratory tests are required. Interested readers may refer to [48] to obtain a complete report on the laboratory setup and measurements of transformer winding capacitances. The extracted capacitances are as follows: C1 $= 1.378$ nF, C2 $= 0.002$ nF, C3 $= 3.513$ nF,

Figure 8b presents a comparison between the behavior of the spark gap against lightning overvoltages by making an experimental laboratory test [49] and the result obtained via the DE method simulated in EMTP-RV. As can be seen in Figure 8b, the result of the DE method is in very good agreement with the laboratory test. It is noteworthy to mention that the small peak after the primary peak in the applied impulse is due to the induced voltage on the standard voltage divider of the laboratory as a result of the electromagnetic fields associated with the high voltage impulse [50]. More information regarding the applied impulse and the validation of the spark gap can be found in [45].

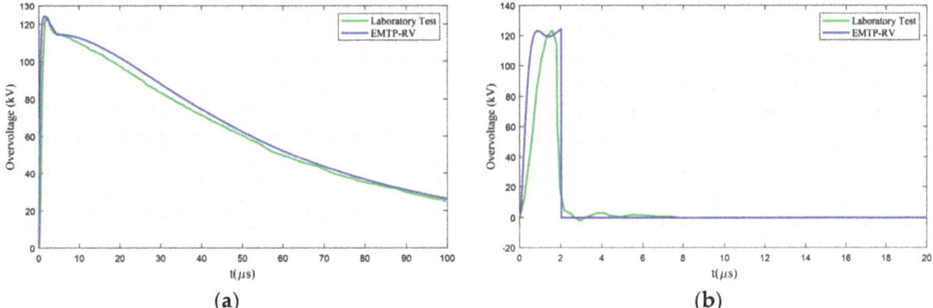

Figure 8. (a) Applied nonstandard 125 kV lightning impulse; (b) Measured overvoltage stress in the presence of spark gap [45].

The performance of the spark gap in a network is verified in Section 3.2 jointly with validating the behavior of the insulator.

3.2. Insulator

Whenever the voltage across the insulator is larger than the flashover voltage calculated by (1), a flashover occurs. That is, under very high overvoltage conditions, an insulator may experience frequent flashovers. Let us investigate different situations for the insulator and see its performance. To do so, the lightning is applied on phase a (see Figure 6) of an intermediate pole adjacent to the transformer pole. Figure 9a–c present the flashover occurrence across the insulators corresponding to phase a, b and c, respectively, while Figure 9d demonstrates the performance of the spark gap in protecting a pole-mounted transformer against lightning impulses. In these figures, the gaps among the bar-like overvoltages stand for the flashover occurrence, and a larger gap means a longer flashover time. Figure 9a–c shows that, by striking the lightning at a pole, frequent flashovers occur across the insulators, and the highest flashover voltage occurs across insulator a (i.e., insulator used for phase a), which is the phase that the lightning strikes. However, from Figure 9d, it can be seen that, for this phase, the spark gap is not triggered. This happens since the time of each flashover across insulator a is comparatively high (see the gaps in Figure 9a); then, the area under the overvoltage curve (which represents the integral part in Equation (4)) is not large enough to trigger the spark gap (see the red bar-like overvoltage waveform in Figure 9d). For phase 2, the spark gap is not triggered since the flashover voltage is always below the onset voltage of primary ionization, which is 112 kV in this paper—see the blue waveform in Figure 9d. However, for insulator c, the situation is different, and the area below the overvoltage curve is large enough to trigger the spark gap and protect the transformer against very high overvoltage stress.

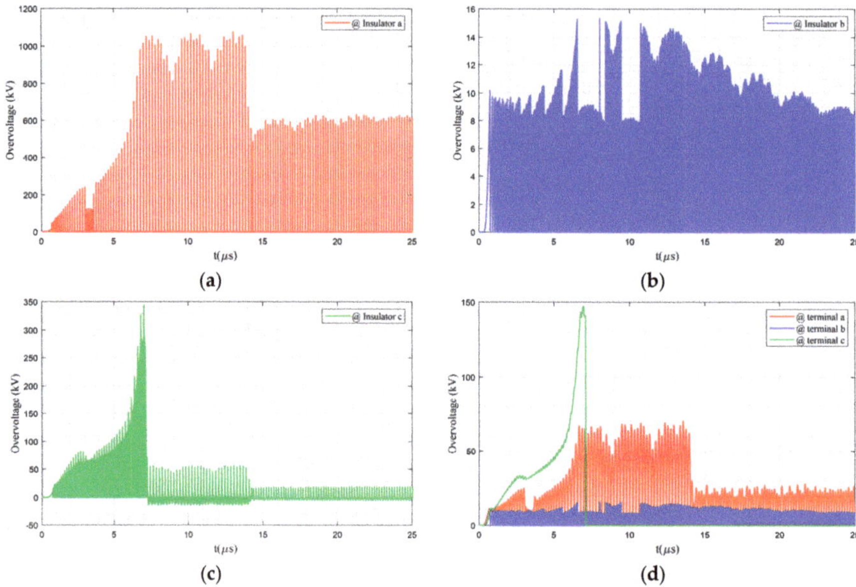

Figure 9. Flashover occurrence across (**a**) insulator a, (**b**) insulator b, and (**c**) insulator c, and (**d**) performance of the spark gap of the same pole against lightning impulses.

3.3. Surge Arrester

The functionality of the utilized surge arrester can be verified by applying the direct lightning and monitoring the residual voltage under the nominal current. The surge arrester model is validated by performing a simulation-based test under the standard 10 kA current impulse (8/20 μs) and comparing the results with laboratory tests, provided in [43]. The obtained result from the simulated model in EMTP-RV shows a residual voltage of 67.32 kV, see Figure 10, which is in resounding agreement with the 67.5 kV residual voltage reported in [43].

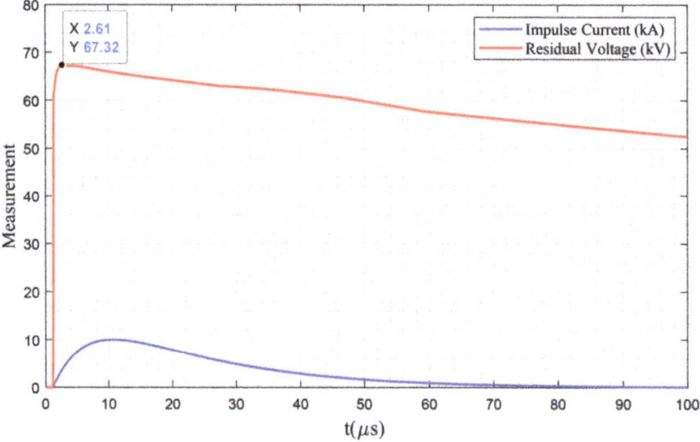

Figure 10. Residual voltage under a 10 kA current impulse (8/20 μs).

Now, let us consider the performance of surge arresters in the sample network. To do so, first, direct lightning is applied on the intermediate pole adjacent to the transformer pole (see Figure 11a), and then direct lightning is applied on the transformer pole (Figure 11b). As can be seen in Figure 11a, the measured overvoltages for all three terminals are below the residual voltage, while, in Figure 11b, when the lightning directly struck the transformer pole, the surge arrester experienced a steep overvoltage with an amplitude of around 80.34 kV. However, immediately, it starts its proper role and the overvoltage is reduced below its residual voltage level, which is a natural behavior of surge arresters against steep lightning strikes [12]. The cases presented in Figure 11a,b verify the proper functionality of surge arresters under normal and critical overvoltage conditions, respectively.

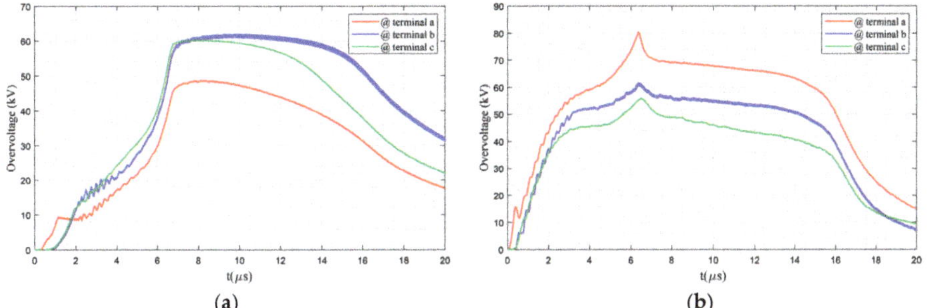

Figure 11. Performance of surge arrester when direct lightning is applied on (**a**) the intermediate pole adjacent to the transformer and (**b**) the transformer pole.

4. Case Studies

This section provides detailed information about the primary assumptions and case studies. To perform in-depth analyses of the protection level of pole-mounted transformers, several case studies are conducted. As mentioned before, two types of wooden poles, namely arsenic- and creosote-impregnated poles, are used for investigations. These poles have been selected since the arsenic-impregnated (AI) pole shows the greatest variations in resistance according to the weather conditions—28 kΩ in November and around 1350 kΩ in August—while the creosote-impregnated (CI) pole experiences the least variations in resistance—580 kΩ in November and 1400 kΩ in August [31]. The primary goals of the following case studies are (a) to define the effectiveness of grounding distance between two successive grounded poles, (b) revealing the role of guy wires in decreasing the amplitude of the lightning overvoltages, as well as (c) investigating the effectiveness of equipping up to two transformer poles with surge arresters on the transient condition of non-surge-arrester-protected transformer poles. In this work, the following assumptions are imposed:

a.   All transformers are protected with spark gaps.
b.   All the transformer poles are grounded, and, like practical cases, in transformer poles, the transformers and the pole have a shared earthing system.
c.   The position of direct lightning has been varied and applied after every three spans, and the amplitudes of the voltages on the primary side of transformers and for all the three phases are measured.
d.   For the sample network, three primary configurations are considered. These configurations are (1) network with wet creosote-impregnated poles, (2) network with wet arsenic-impregnated poles, and (3) network with dry creosote-/arsenic-impregnated poles. It is noteworthy to mention that, since under dry conditions, both the networks with creosote- and arsenic-impregnated poles show relatively similar resistances, only one test is accomplished for them under this weather condition.

In this work, and in order to extract the probability of MV pole-mounted transformers being subjected to overvoltage stress, the following case studies are conducted.

Case 1 Distribution network with grounded transformer poles while other poles are ungrounded. This case performs studies on three primary network configurations (see Section 4, assumption d) where only the transformer poles of the sample distribution network are grounded, and the transformers are protected by the spark gaps.

Case 2 Distribution network with grounded transformer poles and a number of grounded intermediate poles. This case provides a sensitivity analysis to investigate the relationship between the grounding density, i.e., grounding intermediate poles between every two successive transformer poles. To this end, first, only one intermediate pole between two successive transformer poles is grounded; then, the cases with two, three, and four grounded intermediate poles are studied. The primary goal of this case is to investigate the impacts of grounding densities (number of grounded intermediate poles) on mitigating the overvoltage stress on the MV terminals of transformers.

Case 3 Intermediate poles with a guy wire. This case investigates the impacts of considering guy wires on the probabilities of transformers experiencing different stress levels. Additionally, the impact of joint consideration of grounded intermediate poles and guy wires is studied. The setting for this sub-configuration is constructed such that all non-grounded intermediate poles are supported by guy wire.

Case 4 Considering surge arrester for some transformers. This case investigates the impacts of surge arresters on enhancing the protection level. In this case, two different sub-configurations are performed considering two and four surge-arrester-protected transformers.

In this work, the probability of overvoltage stress on the primary side of the transformer, i.e., the probability that the overvoltage amplitude, on at least one phase of the transformer, exceeds the critical overvoltage, is calculated by (5).

$$\pi_{V \geq V_{cr}} = \frac{\sum\limits_{i \in \Lambda_{app}, V_i \geq V_{cr}} n_i}{N_{tr} \cdot N_{app}} \tag{5}$$

where $\pi_{V \geq V_{cr}}$ is the probability of overvoltage stress on the primary side; V_i is the highest voltage among the three phases measured on the primary side; V_{cr} is the critical voltage; Λ_{app} is the set of poles on which the direct lightning is applied (in this paper, $\Lambda_{app} = \{1, 4, 7, \ldots, 82\}$); n_i is the number of transformers in which at least one of the phases exceeds the critical voltage under applied lightning on pole i; N_{tr} is the total number of pole-mounted transformers (in this work, 17), and N_{app} is the total number of poles under direct lightning—in this work, $card(\Lambda_{app}) = 28$.

5. Simulation Results and Discussion

It is assumed that four different stress levels exist for the transformer, although the transformer is essentially under stress if the voltage at the MV side of the transformer reaches above 125 kV, which is one of the standard impulse withstand voltages used for testing the transformer insulation [51] and is applied for the understudy transformer [52]. Therefore, to provide adequate information, besides studying the probability of overvoltage amplitudes above 125 kV, the probabilities of transformers experiencing overvoltage amplitudes above 150 kV (to study the severe cases), above 85 kV, as well as above 30 kV are studied.

5.1. Case 1. Distribution Network with Grounded Transformer Poles

In the MV distribution network, to ensure electrical safety, the transformer tanks are grounded. Therefore, this case is considered as a benchmark for the sake of comparison.

Tables 2–4 present the probabilities of transformers being under four different overvoltage stress levels (above 150 kV, 125 kV, 85 kV, and 30 kV) respectively for three network configurations, namely network with wet creosote-impregnated, wet arsenic-impregnated, and dry poles.

Table 2. Number of transformers with at least one phase reaching the critical overvoltage level. Case 1, wet creosote-impregnated poles and grounded transformer poles.

Applied at Pole	#1	#4	#7	#10	#13	#16	#19	#22	#25	#28	#31	#34	#37	#40
$V \geq 150$ kV		2	1	1	2		2	1	1	2		2		1
$V \geq 125$ kV		2	1	1	2		2	1	1	3		2	1	1
$V \geq 85$ kV		2	1	1	2		2	2	2	3		2	2	1
$V \geq 30$ kV	1	2	3	2	2	1	2	2	2	3	1	2	2	2

Applied at Pole	#43	#46	#49	#52	#55	#58	#61	#64	#67	#70	#73	#76	#79	#82
$V \geq 150$ kV	2		2	1	1	2		2	1	1	2		2	1
$V \geq 125$ kV	2		2	1	1	2		2	1	1	2		2	1
$V \geq 85$ kV	2		2	2	2	2		2	2	2	2		2	1
$V \geq 30$ kV	2	1	2	2	2	2	1	2	2	2	2	1	2	2

Table 3. Number of transformers with at least one phase reaching the critical overvoltage level. Case 1, wet arsenic-impregnated poles and grounded transformer poles.

Applied at Pole	#1	#4	#7	#10	#13	#16	#19	#22	#25	#28	#31	#34	#37	#40
$V \geq 150$ kV		2	1	1	2		2	1	1	2		2		1
$V \geq 125$ kV		2	1	1	2		2	1	1	2		2	1	1
$V \geq 85$ kV		2	2	2	2		2	2	2	2		2	2	2
$V \geq 30$ kV	1	2	3	2	2	1	2	2	2	2	1	2	2	2

Applied at Pole	#43	#46	#49	#52	#55	#58	#61	#64	#67	#70	#73	#76	#79	#82
$V \geq 150$ kV	2		2	1	1	2		2	1	1	2		2	1
$V \geq 125$ kV	2		2	1	1	2		2	1	1	2		2	1
$V \geq 85$ kV	2		2	2	2	2		2	2	2	2		2	1
$V \geq 30$ kV	2	1	2	2	2	2	1	2	2	2	2	1	2	2

Table 4. Number of transformers with at least one phase reaching the critical overvoltage level. Case 1, dry poles and grounded transformer poles.

Applied at Pole	#1	#4	#7	#10	#13	#16	#19	#22	#25	#28	#31	#34	#37	#40
$V \geq 150$ kV		2	1	1	2		2	1	1	2		2	1	1
$V \geq 125$ kV		2	1	1	2		2	1	1	2		2	1	1
$V \geq 85$ kV		2	2	2	2		2	2	2	2		2	2	2
$V \geq 30$ kV	1	2	2	2	2	1	2	2	2	2	1	2	2	2

Applied at Pole	#43	#46	#49	#52	#55	#58	#61	#64	#67	#70	#73	#76	#79	#82
$V \geq 150$ kV	2		2	1	1	2		2	1	1	2		2	1
$V \geq 125$ kV	2		2	1	1	2		2	1	1	2		2	1
$V \geq 85$ kV	2		2	?	2	2		2	2	2	2		2	1
$V \geq 30$ kV	2	1	2	2	2	2	1	2	2	2	2	1	2	1

Results presented in Tables 2–4 show that, although all the transformer poles are grounded, under direct lightning conditions, in several tests (applying direct lightning on different poles), one or two transformers experience very high overvoltage stress (above 150 kV) while, in all tests, at least one transformer is subjected to overvoltage stress above 30 kV. Additionally, it shows that the number of transformers under high overvoltage stress (above 125 kV) is the same as very high overvoltage stress, except for the network configuration with wet creosote-impregnated poles: when the lightning directly strikes pole #28, three transformers experience overvoltage stress above 125 kV while only two transformers experience overvoltage stress above 150 kV.

Figure 12 depicts the probabilities of transformers being subjected to different levels of overvoltage stress. As can be seen, in this case, the probabilities for the different system configurations (wet creosote, wet arsenic, and dry poles) are almost the same. The main reason is that the intermediate poles (poles between two successive transformer poles) are ungrounded, and on the other hand, the dielectric strength of 7 m pole under the wet and dry condition and with both the impregnation materials (creosote or arsenic) is large enough to prevent any flashover. That is, any lightning strike on one of the intermediate poles travels toward the nearest grounded pole (which is the transformer pole in this case), and a flashover on the insulator occurs. It can be seen from Tables 2–4 that, in some tests, only one transformer experiences low overvoltage stress (above 30 kV). This happens when the direct lightning strikes the transformer poles (i.e., poles #1, #16, #31, #46, #61, and #76), and, according to the combination of dielectric strength (see Section 2.2), flashovers on the insulators are discharged via the grounded cross arm.

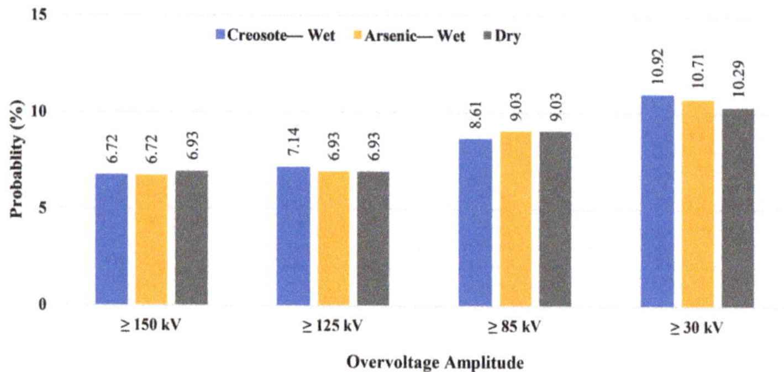

Figure 12. Comparison among the probabilities of transformer overvoltage stress, Case 1 (only transformer poles grounded).

Since this case served as the benchmark, the overvoltage curves at the three terminals of the transformer mounted at pole # 36 are captured in Figure 13. In this figure, the lightning is applied on the adjacent pole of the network with dry poles. As can be seen from this figure, the overvoltage stress at terminal c is much higher than the other two terminals; therefore, for the sake of clarity and not repeating similar concluding remarks, in case 2, case 3, and case 4, only the overvoltage stress at terminal c of this transformer will be considered as the basis for providing in-depth analysis.

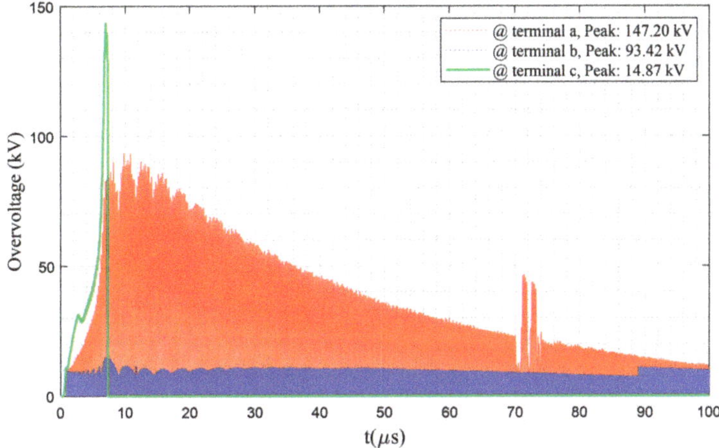

Figure 13. Overvoltage stress at three terminals of transformer mounted at pole #36; Network with dry poles under a lightning strike on pole #35, Case 1.

5.2. Case 2. Distribution Network with Grounded Intermediate Poles

In this case, the three primary network configurations (see Section 4, assumption d) are investigated under several sub-configurations by considering different grounding densities as follows.

C2.1. Grounding one intermediate pole between two successive grounded transformer poles, i.e., poles #3, #8, #13, and so on are grounded.

C2.2. Grounding two intermediate poles between two successive grounded transformer poles, i.e., poles (#3, #4), (#8, #9), (#13, #14), and so on are grounded.

C2.3. Grounding three intermediate poles between two successive grounded transformer poles, i.e., poles (#2, #3, #4), (#7, #8, #9), (#12, #13, #14), and so on are grounded.

C2.4. Grounding all intermediate poles between two successive grounded transformer poles, i.e., poles (#2, #3, #4, #5), (#7, #8, #9, #10), (#12, #13, #14, #15), and so on are grounded.

This case is used to deduce the impact of grounding densities (number of grounded intermediate poles between two successive grounded poles) on the amplitude of lightning overvoltage as well as on the probabilities of transformers experiencing different overvoltage stress levels at their MV terminals.

Tables 5 and 6 respectively stand for the results of grounding one intermediate pole between two successive transformer poles for the network with wet creosote- and wet arsenic-impregnated poles against direct lightning strikes, while Table 7 presents the results for the dry condition.

By comparing Tables 5–7, respectively, with Tables 2–4, one significant difference is that the number of transformers under severe overvoltage stress (overvoltage with an amplitude above 125 kV and 150 kV) has been decreased. On the other hand, a comparison among Tables 5–7 shows that by applying direct lightning on different poles, the numbers of transformers experiencing an overvoltage stress level are almost the same for similar tests, which can be justified according to the large-enough dielectric strength of the 7 m wood poles under different conditions that prevents any flashover and fast overvoltage discharge to the ground. To visualize such similarities, Figure 14 depicts the probabilities of overvoltage stress on the MV terminals of transformers under different network configurations. As can be seen from this figure, the probabilities of transformers experiencing overvoltage amplitude over 150 kV, 125 kV, and 85 kV are exactly the same for all three network configurations, while, for overvoltage amplitude over 30 kV, the network with wet arsenic-impregnated poles, compared with the other two network configurations, shows slightly better performance (with 0.42% lower probability).

Table 5. Number of transformers with at least one phase reaching the critical overvoltage level. Case 2, wet creosote-impregnated poles and grounding one intermediate pole between two transformer poles.

Applied at Pole	#1	#4	#7	#10	#13	#16	#19	#22	#25	#28	#31	#34	#37	#40
$V \geq 150\,\mathrm{kV}$				1					1					1
$V \geq 125\,\mathrm{kV}$				1					1					1
$V \geq 85\,\mathrm{kV}$		1	1	1			1	1	1			1	1	1
$V \geq 30\,\mathrm{kV}$	1	2	2	2	2	1	2	2	2	2	1	2	2	2
Applied at Pole	#43	#46	#49	#52	#55	#58	#61	#64	#67	#70	#73	#76	#79	#82
$V \geq 150\,\mathrm{kV}$					1					1				1
$V \geq 125\,\mathrm{kV}$					1					1			1	1
$V \geq 85\,\mathrm{kV}$			1	1	1			1	1	1			1	1
$V \geq 30\,\mathrm{kV}$	2	1	2	2	2	2	1	2	2	2	2	1	2	1

Table 6. Number of transformers with at least one phase reaching the critical overvoltage level. Case 2, wet arsenic-impregnated poles and grounding one intermediate pole between two transformer poles.

Applied at Pole	#1	#4	#7	#10	#13	#16	#19	#22	#25	#28	#31	#34	#37	#40
$V \geq 150\,\mathrm{kV}$				1					1					1
$V \geq 125\,\mathrm{kV}$				1					1					1
$V \geq 85\,\mathrm{kV}$		1	1	1			1	1	1			1	1	1
$V \geq 30\,\mathrm{kV}$	1	2	2	2	2	1	2	2	2	2	1	2	2	1
Applied at Pole	#43	#46	#49	#52	#55	#58	#61	#64	#67	#70	#73	#76	#79	#82
$V \geq 150\,\mathrm{kV}$					1					1				1
$V \geq 125\,\mathrm{kV}$					1					1			1	1
$V \geq 85\,\mathrm{kV}$			1	1	1			1	1	1			1	1
$V \geq 30\,\mathrm{kV}$	2	1	2	2	1	2	1	2	2	2	2	1	2	1

Table 7. Number of transformers with at least one phase reaching the critical overvoltage level. Case 2, dry poles and grounding one intermediate pole between two transformer poles.

Applied at Pole	#1	#4	#7	#10	#13	#16	#19	#22	#25	#28	#31	#34	#37	#40
$V \geq 150\,\mathrm{kV}$				1					1					1
$V \geq 125\,\mathrm{kV}$				1					1					1
$V \geq 85\,\mathrm{kV}$		1	1	1			1	1	1			1	1	1
$V \geq 30\,\mathrm{kV}$	1	2	2	2	2	1	2	2	2	2	1	2	2	2
Applied at Pole	#43	#46	#49	#52	#55	#58	#61	#64	#67	#70	#73	#76	#79	#82
$V \geq 150\,\mathrm{kV}$					1					1				1
$V \geq 125\,\mathrm{kV}$					1					1			1	1
$V \geq 85\,\mathrm{kV}$			1	1	1			1	1	1			1	1
$V \geq 30\,\mathrm{kV}$	2	1	2	2	2	2	1	2	2	2	2	1	2	1

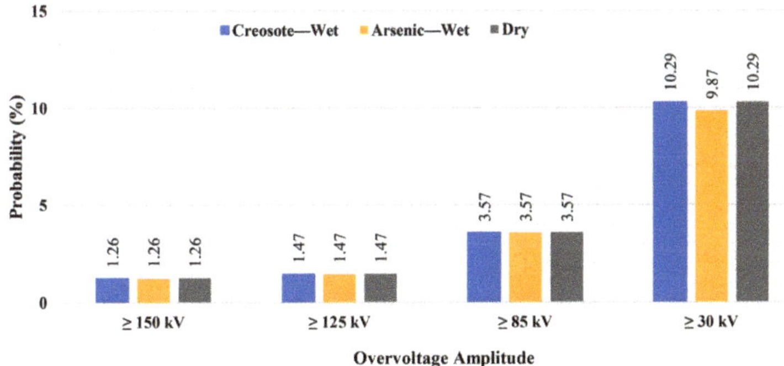

Figure 14. Comparison among the probabilities of transformer overvoltage tension under three network configurations. Case 2, grounding one intermediate pole between two transformer poles.

For the other three sub-configurations (grounding two, three, and four intermediate poles between every two successive transformer poles), a similar situation occurs, i.e., the probabilities of the transformer being under different overvoltage stress levels are almost the same for each individual sub-configuration. To elaborate on the impact of grounding density on the mitigation of overvoltage stress, four overvoltage stress levels are depicted separately in Figures 15–18.

Figure 15 presents a comparison among the probabilities of transformers being under overvoltage stress above 150 kV for three different network configurations (see Section 4, assumption d) due to the effects of different grounding densities, i.e., no grounded intermediate pole (only transformer poles (Only Tr) are grounded) and grounding intermediate poles with different densities (one to four poles). As can be seen in this figure, from the condition with no grounded intermediate poles to grounding only one intermediate pole between two successive grounded transformer poles, the probabilities of transformers being under very high stress (above 150 kV) are considerably decreased by around 5.46%, 5.46%, and 5.67%, respectively, for the wet creosote-impregnated, wet arsenic-impregnated, and dry poles. From one grounded intermediate pole to two grounded intermediate poles, in all configurations, the probabilities decreased by around 1.05%, while by grounding three intermediate poles, the probabilities reach zero, which is the ideal condition for operating a system located in areas with a large number of lightning strikes per year. Moreover, it can be seen that, if a system operator wants to keep the risk of very high overvoltage stress on the MV terminals of the transformers below 2%, only one intermediate pole needs to be grounded; meanwhile, to keep the risk below 1% or eliminate such overvoltage stress, two and three intermediate poles need to be grounded, respectively.

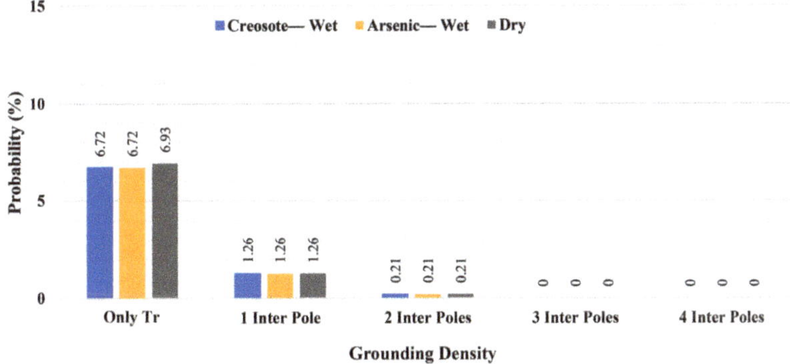

Figure 15. Comparison among the probabilities of transformers experiencing overvoltage stress above 150 kV under different grounding densities, Case 2.

Figure 16 presents a comparison among the probabilities of overvoltage stress higher than 125 kV for three different network configurations according to the effects of different sub-configurations (grounding densities; see Section 5.2, first paragraph). Comparing Figure 16 with Figure 15 shows that there is no significant difference between the trends, where the highest risk belongs to the case in which only the transformer poles are grounded, and the best performance belongs to the case in which at least three intermediate poles are grounded. Moreover, by comparing Figure 16 with Figure 15, it can be seen that, for different network configurations in case 1 (grounding only the transformer poles) and sub-configuration C2.1, the probability of transformers being under overvoltage stress above 125 kV is slightly greater than or equal to the probability of transformers being under overvoltage stress above 150 kV. That is, for case 1, the probabilities in Figure 16 for wet creosote- and wet arsenic-impregnated poles are 0.42% and 0.21% higher than the probabilities in Figure 15, while for dry poles, the probabilities are equal. For case 2 and sub-configuration C2.1, for all network configurations, the probabilities in Figure 16 are 0.21% higher than the probabilities in Figure 15. Consequently, if a system operator is willing to keep the risk of high overvoltage stress below 2% and 1%, one and two intermediate poles need to be grounded, respectively; meanwhile, to eliminate such overvoltage stress, three intermediate poles need to be grounded.

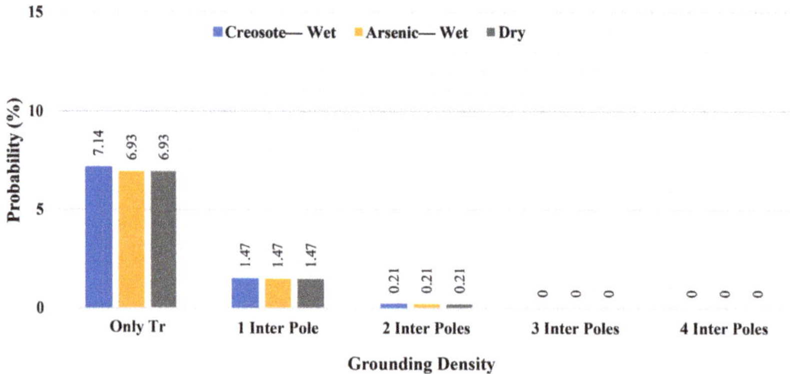

Figure 16. Comparison among the probabilities of transformers experiencing overvoltage stress above 125 kV under different grounding densities, Case 2.

Figure 17 compares the impacts of the grounding densities of the intermediate poles on mitigating the overvoltage stress (above 85 kV) at the MV terminals of transformers for three primary network configurations. In this figure, compared with the very high overvoltage stress and high voltage stress presented in Figures 15 and 16, respectively, the probabilities of the transformer experiencing overvoltage stress are slightly higher. However, the main difference is in the mitigation trend, where, for the overvoltage stresses above 150 kV and 125 kV, by grounding one intermediate pole between every two successive transformer poles, the probability of the transformer experiencing overvoltage stress decreased below 2%, while for this overvoltage level to keep the probabilities below 2%, at least three intermediate poles need to be grounded. On the other hand, by grounding all intermediate poles, the probability of transformers experiencing overvoltage stress above 85 kV is 0%, i.e., there is no risk of such overvoltage stress at the MV terminals of the pole-mounted transformers.

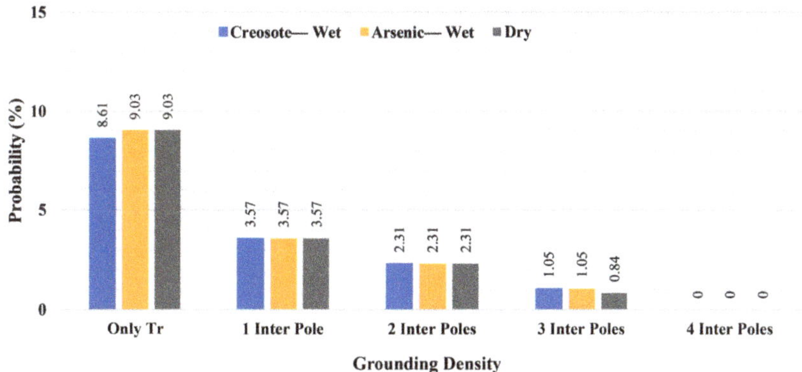

Figure 17. Comparison among the probabilities of transformers overvoltage stress above 85 kV under different grounding conditions, Case 2.

Figure 18 presents a comparison among the probabilities of overvoltage stress above 30 kV for three different network configurations due to the impacts of different sub-configurations (grounding densities). As can be seen from this figure, by grounding one intermediate pole, the highest impact is for the network with arsenic-impregnated poles, which experienced around a 0.84% decrease in the probability of transformers experiencing overvoltage tension above 30 kV. Moreover, it can be seen that by grounding all intermediate poles between every two successive transformer poles, the probabilities are decreased to a great extent; for the network with wet creosote, wet arsenic, and dry poles, compared with Case 1, decrements of around 4.62%, 4.83%, and 3.78% are obtained, respectively. In Figure 18, unlike the results obtained for other overvoltage stress levels (above 150 kV in Figure 15, 125 kV in Figure 16, and 85 kV in Figure 17), grounding intermediate poles does not have a significant effect on mitigating the overvoltage stress above 30 kV. Moreover, for the configuration with four grounded intermediate poles in which no transformer experiences overvoltage stress above 85 kV (see Figure 17), still 6.3%, 5.88%, and 6.51% of transformers in the network with wet creosote, wet arsenic, and dry poles respectively experience overvoltage stress between 30 kV and 85 kV (see Figure 18). Nevertheless, it is noteworthy to mention that, due to the internal insulation of the medium voltage transformers with the basic insulation level above 125 kV [51], such overvoltage stress does not result in immediate failure and its cumulative negative effect over time may be negligible, though a concrete conclusion on this topic requires profound investigation.

Figure 19 provides a comparison among the overvoltage stress at terminal c of the transformer, mounted at pole #36, under different grounding densities for the network with dry poles. The blue curve, which represents case 1 (only transformer poles are grounded), shows that the overvoltage is so high that the spark gap is triggered, and a service interruption occurs. By grounding only one intermediate pole, the amplitude of the overvoltage at terminal c is decreased by around 10.73% (15.79), and such unwanted service interruption is not happening anymore, while the overvoltage stress is still above 125 kV. Additionally, it can be deduced that, as the number of grounded intermediate poles increases, the amplitude of overvoltage decreases such that by grounding all intermediate poles, the overvoltage amplitude is decreased to 40.27 kV, which is a significant enhancement.

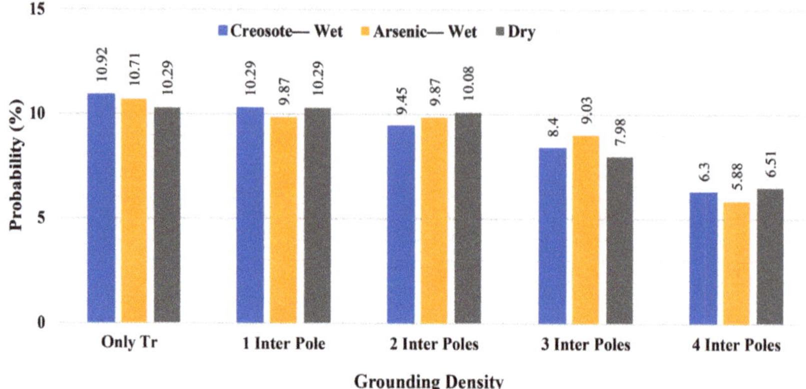

Figure 18. Comparison among the probabilities of transformer overvoltage stress above 30 kV under different grounding conditions, Case 2.

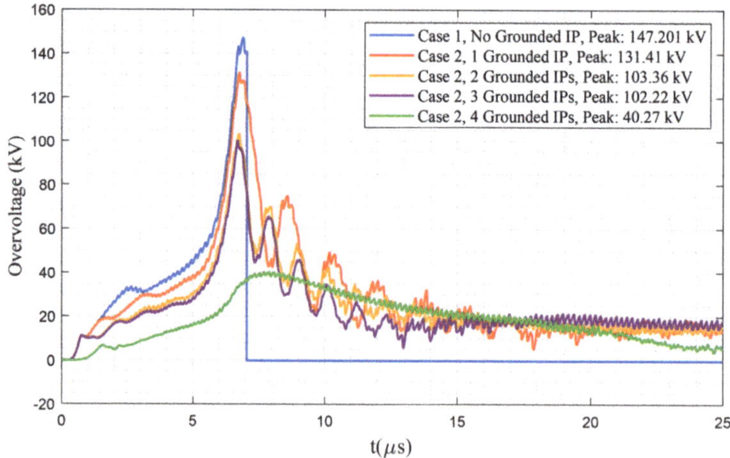

Figure 19. Comparison among the overvoltage stress at terminal c of the transformer mounted at pole #36, network with dry poles under a lightning strike on pole #35, Case 2.

5.3. Case 3. Transformer Poles with Guy Wire

In this case, the three primary network configurations (see Section 4, assumption d) are investigated under four different sub-configurations by considering various combinations of grounding densities and guy wires. These sub-configurations are as follows.

C3.1. Four guy wires (4 GWs): in this sub-configuration, all four intermediate poles between every two successive grounded transformer poles are supported by guy wires.

C3.2. Three guy wires and one grounded intermediate pole (3 GWs + 1 IP): in this sub-configuration, three intermediate poles between every two successive grounded transformer poles are supported by guy wires and the other intermediate pole is grounded. That is, intermediate poles #3, #8, #13, and so on are grounded, while the non-grounded intermediate poles are supported by guy wires.

C3.3. Two guy wires and two grounded intermediate poles (2 GWs + 2 IPs): in this sub-configuration, two intermediate poles between every two successive grounded transformer poles are supported by guy wires and the other two intermediate poles are grounded. That is, intermediate poles (#3,

#4), (#8, #9), (#13, #14), and so on are grounded, while the non-grounded intermediate poles are supported by guy wires.

C3.4. One guy wire and three grounded intermediate poles (1 GW + 3 IPs): in this sub-configuration, one intermediate pole between every two successive grounded transformer poles is supported by guy wires and the other intermediate poles are grounded. That is, intermediate poles (#2, #3, #4), (#7, #8, #9), (#12, #13, #14), and so on are grounded, while the non-grounded intermediate poles are supported by guy wires.

This case is used to evaluate the role of a guy wire on mitigating the overvoltage stress level of transformers against direct lightning phenomena. In other words, this case investigates the impact of guy wire on the amplitude of lightning overvoltage as well as on the probabilities of transformers experiencing different overvoltage stress levels on their MV terminals.

Tables 8–10 present the number of transformers experiencing overvoltage stress after striking direct lightning on different poles of the sample distribution network under three primary configurations, such as the network with wet creosote-impregnated, wet arsenic-impregnated, and dry poles, respectively, while all intermediate poles are supported by guy wires (see sub-configuration C3.1).

Table 8. Number of transformers with at least one phase reaching the critical overvoltage level. Case 3, wet creosote-impregnated poles, sub-configuration C3.1.

Applied at Pole	#1	#4	#7	#10	#13	#16	#19	#22	#25	#28	#31	#34	#37	#40
$V \geq 150\,kV$		2	1	1	1		1	1	1	1				1
$V \geq 125\,kV$		2	1	1	2		2	1	1	2		2	1	1
$V \geq 85\,kV$		2	2	2	2		2	2	2	2		2	2	1
$V \geq 30\,kV$	1	2	2	2	2	1	2	2	2	2	1	2	2	2
Applied at Pole	#43	#46	#49	#52	#55	#58	#61	#64	#67	#70	#73	#76	#79	#82
$V \geq 150\,kV$			1	1	1	1		1	1	1	1		1	1
$V \geq 125\,kV$	2		2	1	1	2		2	1	1	2		2	1
$V \geq 85\,kV$	2		2	1	1	2		2	1	2	2		2	1
$V \geq 30\,kV$	2	1	2	2	2	2	1	2	2	2	2	1	2	2

Table 9. Number of transformers with at least one phase reaching the critical overvoltage level. Case 3, wet arsenic-impregnated poles, sub-configuration C3.1.

Applied at Pole	#1	#4	#7	#10	#13	#16	#19	#22	#25	#28	#31	#34	#37	#40
$V \geq 150\,kV$		1	1	1	1		1	1	1	1			1	1
$V \geq 125\,kV$		1	1	1	2		2	1	1	2		2	1	1
$V \geq 85\,kV$		1	1	1	2		2	1	2	2		2	1	1
$V \geq 30\,kV$	1	3	2	2	2	1	2	2	2	2	1	2	2	2
Applied at Pole	#43	#46	#49	#52	#55	#58	#61	#64	#67	#70	#73	#76	#79	#82
$V \geq 150\,kV$	1		1	1	1	1		1	1	1	1		1	1
$V \geq 125\,kV$	2		1	1	1	2		2	1	1	2		2	1
$V \geq 85\,kV$	2		1	1	1	2		2	1	1	2		2	1
$V \geq 30\,kV$	2	1	2	2	2	2	1	2	2	2	2	1	2	2

Table 10. Number of transformers with at least one phase reaching the critical overvoltage level. Case 3, dry poles, sub-configuration C3.1.

Applied at Pole	#1	#4	#7	#10	#13	#16	#19	#22	#25	#28	#31	#34	#37	#40
$V \geq 150$ kV		2	1	1	1		1	1	1	1				1
$V \geq 125$ kV		2	1	1	2		2	1	1	2		2	1	1
$V \geq 85$ kV		2	2	2	2		2	2	2	2		2	2	1
$V \geq 30$ kV	1	2	3	2	2	1	2	2	2	2	1	2	2	2
Applied at Pole	#43	#46	#49	#52	#55	#58	#61	#64	#67	#70	#73	#76	#79	#82
$V \geq 150$ kV	1		1	1	1	1		1	1	1	1		1	1
$V \geq 125$ kV	2		2	1	1	2		2	1	1	2		2	1
$V \geq 85$ kV	2		2	2	2	2		2	2	2	2		2	1
$V \geq 30$ kV	2	1	2	2	2	2	1	2	2	2	2	1	2	2

By comparing Table 8 with Table 2, it can be seen that supporting all intermediate poles by guy wires results in a reasonable decrease in the number of transformers experiencing overvoltage stress above 150 kV. When the poles are supported by guy wires and the lightning is applied on poles #13, #19, #28, #43, #49, #58, #64, #73, and #79, compared with the case without guy wires (in Table 2), the number of transformers experiencing such overvoltage stress decreases by one, while applying the lightning on pole #34 shows even better results, where no transformer experiences such overvoltage stress, which is two transformers less than the case without guy wire (in Table 2). Moreover, comparing Tables 9 and 10 with Tables 3 and 4 shows similar positive impacts of using guy wires on decreasing the overvoltage amplitude at the MV terminals of the transformer. To present the impacts more clearly, the probabilities of overvoltage stress on the MV terminals of transformers for sub-configuration C3.1 and under different network configurations are compared with each other in Figure 20. As can be seen from this figure, for the overvoltage stress with an amplitude above 150 kV (as the most likely stress for causing sever failure during the specified time period), when guy wires support the intermediate poles between two successive transformer poles, the probabilities for all three network configurations, i.e., wet creosote, wet arsenic, and dry poles, are decreased by around 2.52%, 2.31%, and 2.52%, respectively. Moreover, Figure 20 shows that, for other overvoltage stress levels, the guy wires provide some protection against direct lightning strikes, although this is negligible in some sub-configurations.

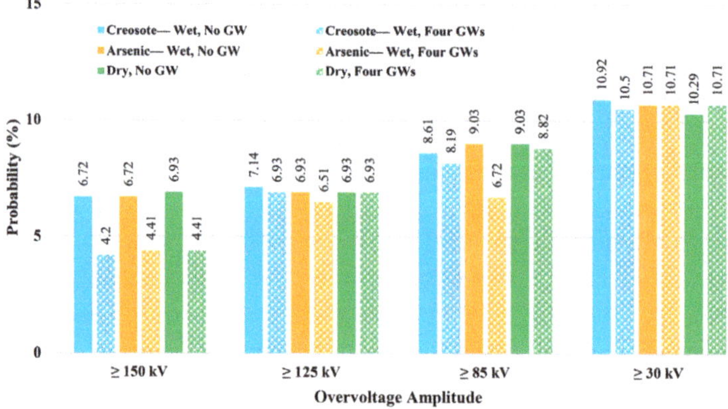

Figure 20. Comparison among the probabilities of transformers experiencing four overvoltage stress levels for three configurations of the network with no guy wire (No GW) and four guy wires (Four GWs).

Another observation in this case is that the guy wires show better protective performance for overvoltage stress with an amplitude above 150 kV than the other overvoltage stress levels. The main reason for such behavior is the occurrence of flashover down the wood pole since, by installing guy wires, the total CFO, which is the combined CFO of the insulators and the wood pole corresponding to the length from cross arm to the guy wire connection point (i.e., 2 m in this paper), is lower than the CFO of the similar case without guy wire (a 7 m wood pole). This situation will be elaborated more by investigating the flashover occurrence on the intermediate poles under three network configurations and for case 1 and sub-configuration C3.1. Tables 2–4 show that when direct lightning is applied on pole #34, for all three network configurations, two transformers on poles #31 and #36 experience overvoltage stress above 150 kV. Therefore, to see the impacts of guy wires, the flashover occurrence on the intermediate poles between transformer poles #31 and #36 (i.e., poles #32, #33, #34, and #35) are studied for case 1 (no guy wire for intermediate poles) and sub-configuration C3.1 (all intermediate poles supported by guy wire). It is worth mentioning that the studies are conducted under the dry condition, in which the wood poles have the highest CFO, e.g., 400 kV/m compared with wet conditions (see Section 2.2).

As mentioned in Section 2.2, the flashover occurrence over the pole is modeled by a voltage-controlled switch. In this paper, to verify the flashover occurrence, the switch status and the current flow through this switch is monitored. In case 1, monitoring the switch status shows that the direct lightning does not result in any flashover down the wood poles, and this is due to the high CFO of the structure (combined CFO of insulators and the 7 m wood pole). However, for sub-configuration C3.1, the situation is different since the combined CFO is lower than the case without considering the guy wires, since the distance from the cross arm to the connection pint of the guy wire is 2 m. Details of flashover occurrence for the intermediate poles of the dry network configuration are depicted in Figure 21.

Figure 21a–d present the current flows through the voltage-controlled switches of four intermediate poles, namely #32, #33, #34, and #35, respectively, due to flashover occurrence as a result of direct lightning strike on pole #34. As can be seen from this figure, for a short period of time, several flashovers frequently occurred down the poles toward the guy wires. It is noteworthy to mention that higher amplitudes of current flow through the switches show that the poles experienced higher overvoltages. Among the aforementioned intermediate poles, pole #34, unlike the other intermediate poles, experienced two discontinuous periods of frequent flashovers because direct lightning was applied on this pole. Moreover, the flashover discharge currents of the intermediate poles show that not only does the highest discharge current belong to pole #34 but the first period of frequent flashovers lasts longer, i.e., it starts at around 5.7 µs and ends at around 7.3 µs. Meanwhile, for pole #33, which has the second widest frequent flashover period, it starts at around 6.2 µs and ends at around 7.2 µs. All in all, as can be seen in Figure 21, when flashover occurs from the cross arm toward the guy wire, the current associated with overvoltage is discharged via guy wires and, consequently, the transformers mounted at poles #31 and #36 in sub-configuration C3.1 experience lower overvoltage amplitude than the same transformers in case 1—see Figure 22. Figure 22 depicts the overvoltage waveform of terminal c of the transformer, i.e., the terminal which is experiencing the highest overvoltage stress, among others. As can be seen from Figure 22a,b that, considering guy wires, the overvoltage amplitudes experienced by the transformers mounted at poles #31 and #36 are 16.99% and 14.97% lower than the overvoltage amplitudes of the case without guy wires. The main reason that the spark gap is not triggered for the transformer can be found in Section 3.2.

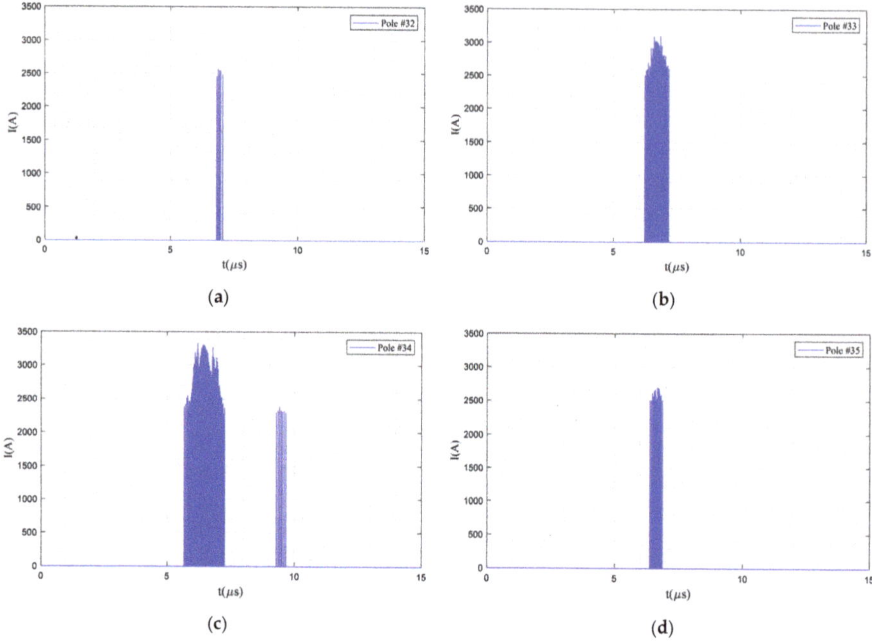

Figure 21. Flashover discharge currents of four intermediate poles supported by guy wires: (**a**) Intermediate Pole #32, (**b**) Intermediate Pole #33, (**c**) Intermediate Pole #34, and (**d**) Intermediate Pole #35; Network with dry poles and direct lightning applied on pole #34.

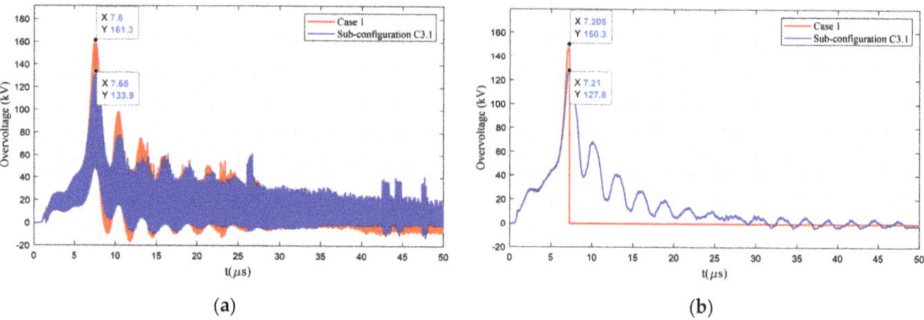

Figure 22. Overvoltage at terminal c of the transformer mounted on (**a**) pole #31, and (**b**) pole #36; Network with dry poles and direct lightning applied on pole #34.

Figure 23 provides a comparison among the probabilities of pole-mounted transformers experiencing four levels of overvoltage stresses before and after considering guy wires and under different grounding densities for the network with wet creosote-impregnated poles. The solid bars stand for the probabilities of under-stress transformers with different grounding densities and without considering guy wires, while the patterned bars represent the probabilities of under-stress transformers when the non-grounded intermediate poles are supported by guy wires. In this figure, for overvoltage amplitudes above 150 kV, when no intermediate pole is grounded, the guy wires show their best performance. That is, the probability of transformers experiencing such overvoltage stress when considering guy wires (blue patterned bar) is decreased by around 2.52%, compared with the similar grounding condition but without guy wires (solid blue bar). By increasing the number of grounded intermediate poles (from blue bars toward black bars), the impact of installing guy wires (at the non-grounded intermediate poles) on mitigating the overvoltage stress is decreased. For example, under overvoltage above 150 kV, when one intermediate pole between all transformer poles is grounded, considering guy wires for non-grounded intermediate poles (patterned orange bar) results in around a 0.84% decrease in the probabilities, compared with similar grounding conditions and without guy wires (solid orange bar). For other grounding conditions of overvoltage stress above 150 kV (green and black bars), the guy wire does not provide any support against lightning overvoltage. This is mainly due to the functionality of grounded poles in the mitigation of high overvoltage stress. Each lightning strike chooses the shortest path with the least dielectric strength toward the ground, and when the number of grounded intermediate poles increases, the discharges occur more easily through grounded poles; this means that the overvoltage amplitude does not reach the CFO of guy wire-supported poles to trigger a flashover. Moreover, a comparison of the probabilities for all overvoltage stress levels shows that the best performance belongs to overvoltage stress above 150 kV, which most likely triggers the flashover discharge.

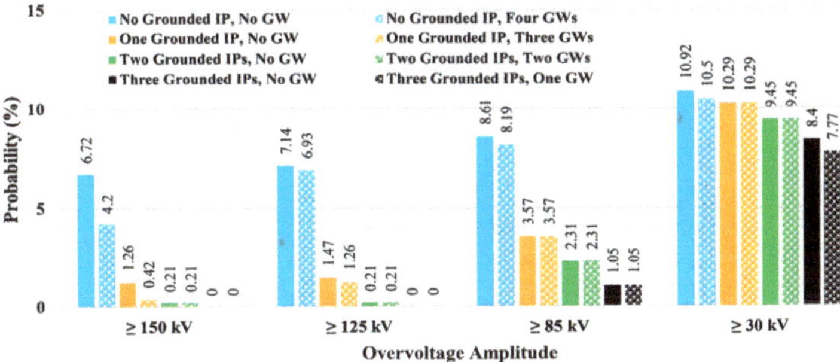

Figure 23. Comparison among the probabilities of transformers experiencing four overvoltage stress levels under different combinations of grounding and guy wire (GW) configurations of intermediate poles (IP); Network with wet creosote-impregnated poles.

Figures 24 and 25 compare the probabilities of pole-mounted transformers experiencing different overvoltage stress levels before and after considering guy wires and under different grounding densities for the network with wet arsenic-impregnated poles and dry poles, respectively. The main difference between using guy wires in the networks with wet creosote-impregnated poles and wet arsenic-impregnated poles is that the wet arsenic configuration has lower CFO. That is, wet arsenic poles require lower overvoltage amplitude than the wet creosote poles to trigger the flashover discharge (see Section 2.2). However, dry poles show stronger dielectric characteristics and, consequently, higher overvoltage amplitude is required for triggering a flashover down the pole toward the guy wire. Comparing the probabilities of overvoltage stress above 150 kV in Figures 24 and 25 with Figure 23 shows that, for sub-configuration C3.1 (patterned blue bars), the guy wire decreases the probabilities in all three different network configurations by more than 2%, which is a considerable performance for such inexpensive devices. For arsenic-impregnated poles and with no grounded intermediate poles (blue bars in Figure 24), the guy wires provide great support against direct lightning even for overvoltage amplitudes above 85 kV where, compared with the same grounding condition in the absence of guy wires (solid blue bar), considering guy wires results in a decrease of around 2.31%.

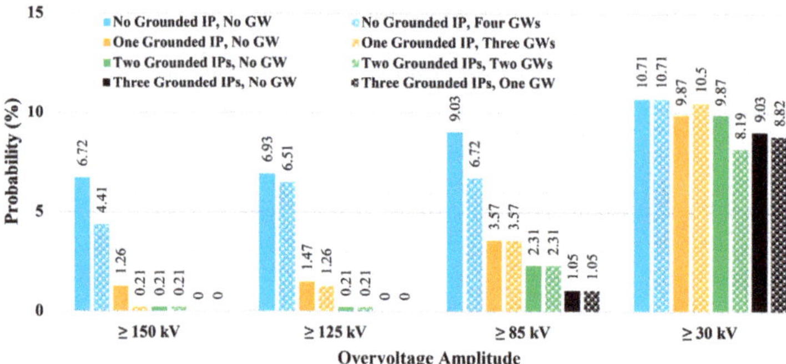

Figure 24. Comparison among the probabilities of transformers experiencing four overvoltage stress levels under different combinations of grounding and guy wire (GW) configurations of intermediate poles (IP); Network with wet arsenic-impregnated poles.

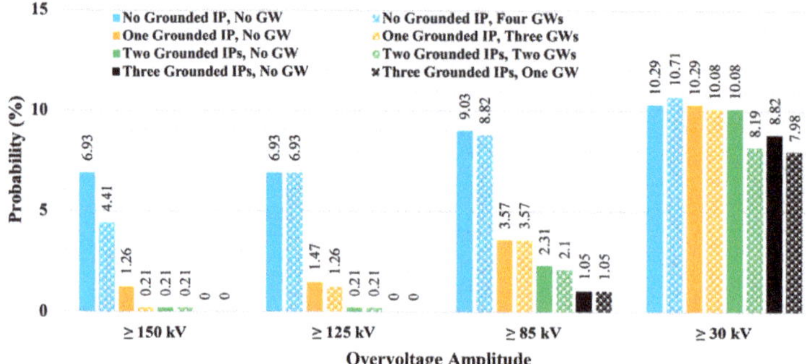

Figure 25. Comparison among the probabilities of transformers experiencing four overvoltage stress levels under different combinations of grounding and guy wire (GW) configurations of intermediate poles (IP); Network with dry poles.

From the results obtained in case 3, it can be deduced that, for overvoltage stress above 150 kV, grounding one intermediate pole and supporting non-grounded poles with guy wires yields almost the same results as grounding two intermediate poles. For example, in Figures 23–25, when only one intermediate pole is grounded and none of the intermediate poles are supported by guy wires, all the probabilities for overvoltage stress above 150 kV are 1.26% (see the solid orange bars), and by grounding one more intermediate pole, the probabilities decrease to 0.21% (see the solid green bars). On the other hand, grounding one intermediate pole and supporting the non-grounded intermediate poles with guy wires shows that, for the network with wet arsenic-impregnated and dry poles (Figures 24 and 25, respectively), the probabilities decrease to 0.21%, which is similar to the case with two grounded intermediate poles. However, for the network with creosote-impregnated poles and one grounded intermediate pole, supporting the non-grounded intermediate poles with guy wires decreases the probability to 0.42%, which is slightly higher than the probability obtained by grounding two intermediate poles.

Figure 26 provides a comparison among the overvoltage stress on terminal c of the transformer at pole #36 under different combinations of grounding densities and guy wires of intermediate poles for the network with dry poles. The red curve, which represents the overvoltage waveform for sub-configuration C3.1, compared with case1 (the blue curve), shows that, although considering four guy wires between every two successive transformer poles decreases the amplitude of overvoltage, it is still high enough to trigger the spark gap and cause a service interruption. However, by supporting one intermediate pole with a guy wire in sub-configuration C3.2 (the orange curve), the amplitude of overvoltage is decreased by around 22.8 kV. As a consequence, the spark gap is not triggered, and no service interruption occurs. Additionally, supporting three intermediate poles with guy wires and grounding the other intermediate pole (sub-configuration C3.2), compared with the case with only one grounded intermediate pole (sub-configuration C2.1, the red curve in Figure 19), results in a considerable enhancement in the overvoltage protection against direct lightning, i.e., 11.67 kV decrease in the overvoltage stress. Similarly, by comparing Figure 26 with Figure 19, it can be deduced that other combinations of guy wire and grounded intermediate pole also reduce the overvoltage stress on the transformer terminal.

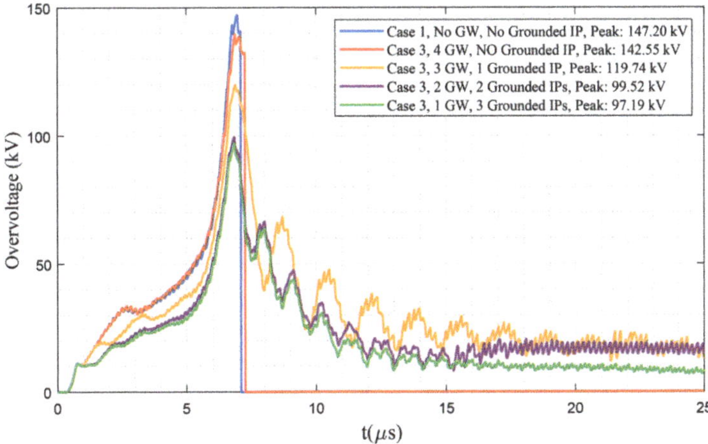

Figure 26. Comparison among the overvoltage stress at terminal c of the transformer mounted at pole #36 under different combinations of grounding and guy wire (GW) configurations of intermediate poles (IP); Network with dry poles under a lightning strike on pole #35.

5.4. Case 4. Considering Surge Arrester for Some Transformers

In this case, the three primary network configurations (see Section 4, assumption d) are investigated under two different sub-configurations by equipping two or four pole-mounted transformers with surge arresters. These sub-configurations are as follows.

C4.1. Equipping two pole-mounted transformers with surge arresters (2 SAs). To do so, the total spans are divided into three. Therefore, the transformer poles are poles #26 and #56. This sub-configuration is studied for all different grounding densities presented in case 2.

C4.2. Equipping four pole-mounted transformers with surge arresters (4 SAs). To do so, the total spans are divided into five. Therefore, the transformers protected by surge arresters are located on poles #16, #36, #56, and #76. This sub-configuration is studied for all different grounding densities presented in case 2.

This case is used to evaluate the impacts of equipping some transformers with surge arresters on the probability of transformers experiencing different overvoltage stress levels. Although the surge arresters are quite effective in protecting the transformers against overvoltage stress [45], their high capital costs prevent equipping all the transformers in a medium voltage network with these devices.

Tables 11–13 present the number of transformers experiencing different overvoltage stress levels after striking direct lightning on different poles of the sample distribution network under three primary configurations, namely the network with wet creosote-impregnated, wet arsenic-impregnated, and dry poles, respectively, where two transformers are equipped with surge arresters and none of the intermediate poles are grounded.

Table 11. Number of transformers with at least one phase reaching the critical overvoltage level. Case 4, wet creosote-impregnated poles, sub-configuration C4.1 with no grounded intermediate poles.

Applied at Pole	#1	#4	#7	#10	#13	#16	#19	#22	#25	#28	#31	#34	#37	#40
$V \geq 150\,kV$		2	1	1	2		2	1			1		1		1
$V \geq 125\,kV$		2	1	1	2		2	1			1		2	1	1
$V \geq 85\,kV$		2	2	2	2		2	1	1	1		2	1	2
$V \geq 30\,kV$	1	2	3	2	2	1	3	3	2	2	2	3	2	2
Applied at Pole	#43	#46	#49	#52	#55	#58	#61	#64	#67	#70	#73	#76	#79	#82
$V \geq 150\,kV$	2		2	1		1		2	1	1	2		2	1
$V \geq 125\,kV$	2		2	1		1		2	1	1	2		2	1
$V \geq 85\,kV$	2		2	1	1	1		2	2	2	2		2	1
$V \geq 30\,kV$	2	1	3	2	2	2	1	3	3	2	2	1	2	2

Table 12. Number of transformers with at least one phase reaching the critical overvoltage level. Case 4, wet arsenic-impregnated poles, sub-configuration C4.1 with no grounded intermediate poles.

Applied at Pole	#1	#4	#7	#10	#13	#16	#19	#22	#25	#28	#31	#34	#37	#40
$V \geq 150\,kV$		2	1	1	2		2	1			1	2		1
$V \geq 125\,kV$		2	1	1	2		2	1			1	2	1	1
$V \geq 85\,kV$		2	2	2	2		2	1	1	1		2	2	2
$V \geq 30\,kV$	1	2	3	2	2	1	3	2	1	2	2	3	2	2
Applied at Pole	#43	#46	#49	#52	#55	#58	#61	#64	#67	#70	#73	#76	#79	#82
$V \geq 150\,kV$	2		2	1		1		2	1	1	2		2	1
$V \geq 125\,kV$	2		2	1		1		2	1	1	2		2	1
$V \geq 85\,kV$	2		2	1	1	1		2	2	2	2		2	1
$V \geq 30\,kV$	2	1	3	2	2	2	1	3	2	2	2	1	2	2

Table 13. Number of transformers with at least one phase reaching the critical overvoltage level. Case 4, dry poles, sub-configuration C4.1 with no grounded intermediate poles.

Applied at Pole	#1	#4	#7	#10	#13	#16	#19	#22	#25	#28	#31	#34	#37	#40
$V \geq 150\,kV$		2	1	1	2		2	1			1	2		1
$V \geq 125\,kV$		2	1	1	2		2	1			1	2	1	1
$V \geq 85\,kV$		2	2	2	2		2	1	1	1		2	1	1
$V \geq 30\,kV$	1	2	3	2	2	1	3	2	2	2	2	3	2	2
Applied at Pole	#43	#46	#49	#52	#55	#58	#61	#64	#67	#70	#73	#76	#79	#82
$V \geq 150\,kV$	2		2	1		1		2	1	1	2		2	1
$V \geq 125\,kV$	2		2	1		1		2	1	1	2		2	1
$V \geq 85\,kV$	2		2	1	1	1		2	2	2	2		2	1
$V \geq 30\,kV$	2	1	3	2	2	2	2	3	2	3	2	1	2	2

By comparing Table 11 with Table 2, it can be seen that, by equipping two transformers (at poles #26 and #56), the number of transformers experiencing overvoltage stress has been decreased only when the lightning is applied on the poles adjacent to these transformers, such as poles #25, #28, #55, and #58. For example, when the lightning strikes directly pole #25, the number of transformers experiencing overvoltage stress above 150 kV and 125 kV is decreased to zero, while, without considering surge arresters, one transformer experiences such overvoltage stress (see Table 2). Also, Table 2 shows that, when the lightning is applied on pole #28, before considering any surge arrester, the numbers of transformers under overvoltage stress above 150 kV and 125 kV are two and three, respectively, while by installing surge arresters, only one transformer experiences such overvoltage stress (see Table 11). Comparing Tables 12 and 13 with Tables 3 and 4 shows a similar positive impact of considering surge arresters on mitigating the overvoltage stress at the MV terminals of some transformers. To present the impacts more clearly, the probabilities of overvoltage stress on MV terminals of transformers for sub-configuration C4.1 and C4.2 under different grounding densities and for three primary network configurations (see Section 4, assumption d) are presented in Figures 27–30. In all of these figures, for each primary configuration of the network, three different conditions are studied, such as without surge arrester (case 1, presented by solid bars), with two surge-arrester-protected transformers (sub-configuration C4.1, presented by patterned bars), and with four surge-arrester-protected transformers (sub-configuration C4.2, presented by diagonal stripes).

Figure 27 presents the probabilities of transformers experiencing overvoltage stress above 150 kV under different grounding densities. As can be seen from this figure, for the case without grounded intermediate poles (only all transformers (Only Tr) grounded), considering two and four transformer poles results in some mitigation in the overvoltage stress. By protecting two transformers with a surge arrester, both the networks with wet creosote-impregnated poles (blue bars) and dry poles (gray bars) show around a 1.05% decrement in the probabilities of transformers experiencing overvoltage stress, while for the network with wet arsenic-impregnated poles, it results in 0.84% decrement. On the other hand, in this figure, the best performance for all network configurations is obtained when four transformers are equipped with surge arresters where, compared with the scenarios without surge arrester (solid bars), the networks with wet arsenic poles and dry poles show a 1.68% decrement in the probabilities and the network with wet creosote poles shows a 1.47% decrement. Figure 27, compared with the similar overvoltage stress level (above 150 kV) in Figures 23–25, shows that considering guy wires for all intermediate poles results in a better outcome than equipping four transformers in the network with surge arresters. Needless to say, and as concluded in case 2, by grounding only one intermediate pole, a great enhancement in the protection level of the transformers is obtained. On the other hand, as can be seen in Figure 27, after grounding one intermediate pole, using surge arresters slightly decreases the probabilities, while grounding two intermediate poles means that the surge arrester does not have a supportive role against direct lightning strikes, and in one scenario (gray diagonal stripes), it negligibly increases the probability of overvoltage stress.

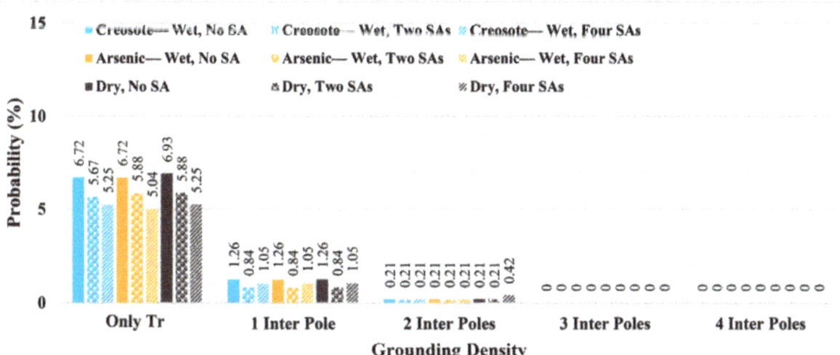

Figure 27. Comparison among the probabilities of transformers experiencing overvoltage stress above 150 kV under different grounding densities of intermediate poles (Inter Pole) and considering two and four surge arresters (SAs); Network with dry poles.

Figures 28 and 29 present the probabilities of transformers experiencing overvoltage stresses above 125 kV and 85 kV, respectively. The probabilities are reported for three primary network configurations under different grounding densities and three different sub-configurations, such as without considering any surge arrester as well as considering two and four transformers with surge arresters. Similar to the overvoltage above 150 kV, under both overvoltage stress levels above 125 kV and 85 kV, the surge arresters are triggered and start protecting the transformers. This can be seen mainly when only all transformers (Only Tr) are grounded and all the intermediate poles remain ungrounded. For example, in Figure 28, after considering two transformers with surge arresters, the probabilities in the Only-Tr sub-configuration decrease by around 1.05%, 0.84%, and 0.84%, respectively, for the network with wet creosote, wet arsenic, and dry poles, while by equipping four transformers by surge arresters, the probabilities are decreased by about 1.68%, 1.47%, and 1.47%, respectively. A similar condition can be observed for the probabilities of transformers experiencing overvoltage stress above 85 kV in Figure 29. Another observation in Figure 29 is related to the protection enhancement for sub-configurations with two and even three grounded intermediate poles that cannot be seen for overvoltage stress above 150 kV (in Figure 27) and above 125 kV (in Figure 28).

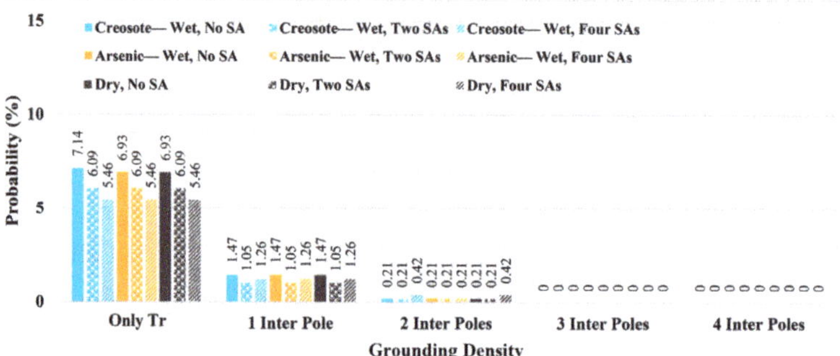

Figure 28. Comparison among the probabilities of transformers experiencing overvoltage stress above 125 kV under different grounding densities of intermediate poles (Inter Pole) and considering two and four surge arresters (SAs); Network with dry poles.

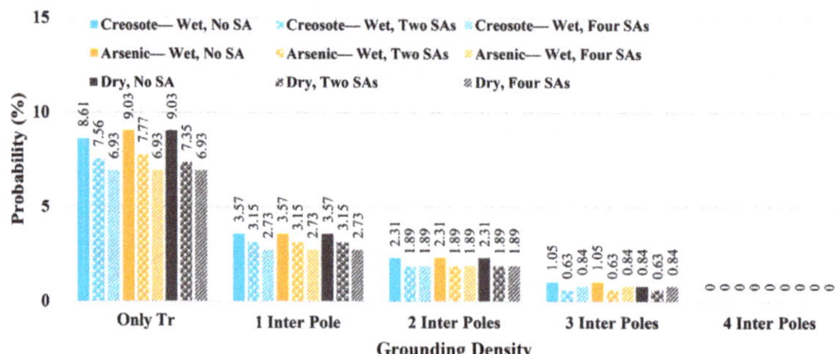

Figure 29. Comparison among the probabilities of transformers experiencing overvoltage stress above 85 kV under different grounding densities of intermediate poles (Inter Pole) and considering two and four surge arresters (SAs); Network with dry poles.

Unlike the three previous overvoltage stress levels, the overvoltage stress above 30 kV shows an unpredictable behavior where considering surge arresters in some scenarios results in increasing the probabilities of transformers experiencing such overvoltage stress. However, this can be justified by considering previous cases. Surge arresters are mainly used to protect the power system equipment against overvoltage phenomena by limiting the overvoltage amplitude or discharging the surge currents associated with surge overvoltages [53]. Therefore, in the sample MV network, these devices are not triggered for overvoltage stress around 30 kV. As a result, and as can be seen in Figures 27–29, the surge arrester is triggered for overvoltage levels above 150 kV, 125 kV, and 85 kV, and the amplitude of overvoltage at the MV terminals of some transformers is decreased to a level below 85 kV, which is accumulated as the probability of an overvoltage stress level above 30 kV.

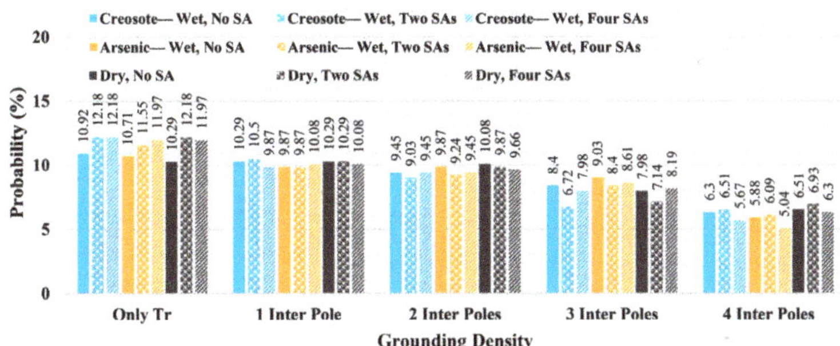

Figure 30. Comparison among the probabilities of transformers experiencing overvoltage stress above 30 kV under different grounding densities of intermediate poles (Inter Pole) and considering two and four surge arresters (SAs); Network with dry poles.

Figure 31 presents a comparison among the overvoltage stress on terminal c of the transformer at pole #36 under different grounding densities for the network with dry poles and considering two surge arrester protected transformers, respectively. By considering two surge arresters, the nearest surge arrester to the transformer pole #36 is placed at pole #26 (see sub-configuration C4.1). From the red curve in Figure 31, which presents the overvoltage waveform under no grounded intermediate pole, it can be seen that the overvoltage amplitude, compared with case 1, is decreased by around 3%, while the spark gap is still triggered. The peak values presented in Figure 31 compared with similar grounding densities in case 2 (see Figure 19) show that the decreases and increases in overvoltage are negligible, and, consequently, if the surge arrester is far from the location of the lightning strike, it might not enhance the network's protection against lightning overvoltages. On the other hand, a comparison with the results of case 3 (see Figure 26) shows that guy wires are much more effective than the impact of a surge arrester placed far from the lightning strike location.

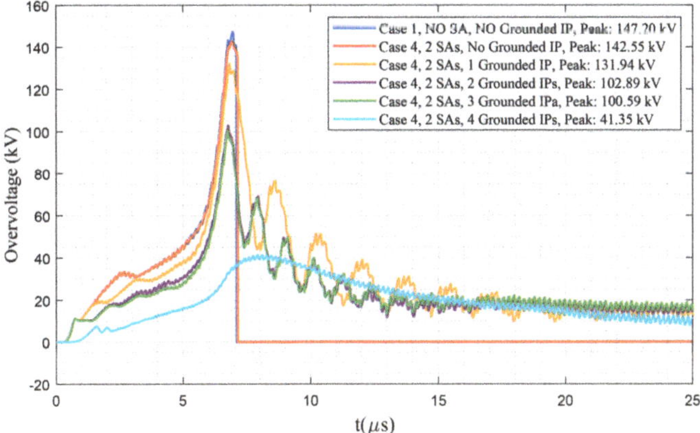

Figure 31. Comparison among the overvoltage stress at terminal c of the transformer mounted at pole #36 under different grounding densities; Network with dry poles considering two surge arrester protected transformers and under a lightning strike on pole #35.

Figure 32 considers similar sub-configurations as Figure 31, but the network has four surge arrester-supported transformers. By considering four surge arresters in the network, one surge arrester is paced at pole # 36 (see sub-configuration C4.2). This case shows a significant difference not only with case 1 but also with similar grounding conditions in case 2, case 3, and case 4 with two surge arresters. As expected, the surge arrester, by absorbing extra energy from the lightning, decreases the overvoltage stress at the terminals of the transformers and consequently shows the best performance among all other sub-configurations. That is, when a surge arrester is placed near the lightning strike, it significantly increases the protection level of the network. In other words, to have such a strong protection level in a network, all the transformers need to be protected by proper surge arresters, which is certainly a costly action.

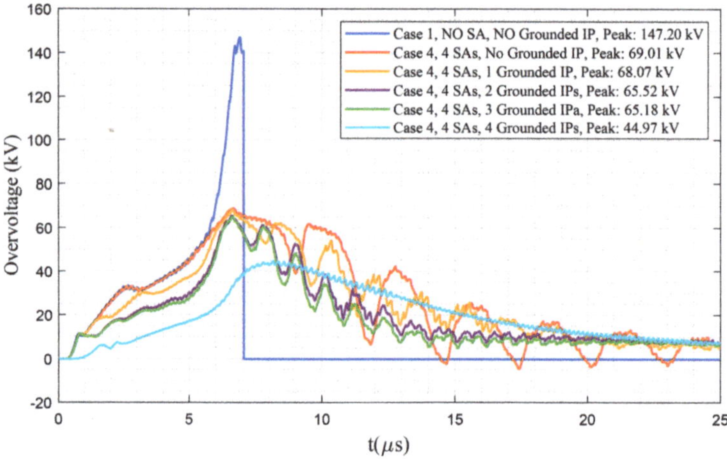

Figure 32. Comparison among the overvoltage stress at terminal c of the transformer mounted at pole #36 under different combinations of grounding and guy wire (GW) configurations of intermediate poles (IP); Network with dry poles considering four surge-arrester-protected transformers under a lightning strike on pole #35.

6. Conclusions

This work has investigated the role of protecting devices which aim to provide optimized protection of medium voltage (MV) distribution transformers against direct lightning strikes on the phase conductor. To this end, the roles of grounding densities of intermediate poles, guy wire, and surge arresters on mitigating the overvoltage stress on the MV terminals of distribution transformers have been thoroughly studied while assuming that all transformers are by default protected by spark gaps and all transformer poles are grounded. It has been shown that grounding the intermediate poles (even only one intermediate pole) can significantly decrease the risk of overvoltage stress on the MV terminals of distribution pole-mounted transformers for all four considered stress levels above 150 kV, 125 kV, 85 kV, and 30 kV. The best outcome is achieved when all poles (transformer and intermediate poles) are grounded; however, this might not be the optimal practice. Additionally, it has been witnessed that supporting intermediate poles with guy wires significantly decreases the probabilities of transformers experiencing high voltage stress. However, when the number of grounded intermediate poles is increased, the performance of guy wires is inferior to the better performance of grounding. In other words, guy wires show better performance when none of the intermediate poles are grounded or the number of grounded intermediate poles does not exceed two poles between every two successive transformer poles. On the other hand, equipping some transformers with surge arresters provides great support only for the adjacent transformers, such that even equipping four transformers out of 17 does not provide the protection that the guy wires provide for the network. That is, its support for the network would be significant when most of the transformers are equipped with this device.

All in all, the obtained results, presented in several tables and figures, as well as the provided interpretations, can serve as a reference table to provide a clear picture for the wooden pole-based system operator to see how much improvement in the protection levels of the transformers in a network can be achieved by constructing grounding systems, utilizing guy wires, and equipping some pole-mounted transformers with surge arrester devices. That is, a system operator can learn the impact of considering the aforementioned protective devices and different system configurations on decreasing the probabilities of overvoltage stress on the medium voltage transformers.

The future works will study the impacts of the grounding densities and guy wire on the probabilities of 1-phase, 2-phase, and 3-phase faults.

Author Contributions: M.P.-K. developed the main idea, performed the simulations, and prepared the original research draft. F.M. provided support for the parameter adjustments of some analytical models and helped with technical verification. M.L. proposed the main idea and supervised the work. All authors have read and agreed to the published version of the manuscript.

Funding: The APC was funded by Aalto University.

Conflicts of Interest: The authors declare no conflict of interest.

References

1. Keitoue, S.; Murat, I.; Filipović-Grčić, B.; Župan, A.; Damjanović, I.; Pavić, I. Lightning caused overvoltages on power transformers recorded by on-line transient overvoltage monitoring system. *J. Energy Energ.* **2018**, *67*, 44–53.

2. Cigre Working Group A2.37. *Transformer Reliability Survey*; Technical Brochure NO. 642: Paris, France, 2015.

3. Tenbohlen, S.; Jagers, J.; Vahidi, F. Standardized survey of transformer reliability: On behalf of CIGRE WG A2.37. In Proceedings of the 2017 International Symposium on Electrical Insulating Materials (ISEIM), Toyohashi, Japan, 11–15 September 2017; pp. 593–596.

4. Furgał, J.; Kuniewski, M.; Pająk, P. Analysis of internal overvoltages in transformer windings during transients in electrical networks. *Energies* **2020**, *13*, 2644. [CrossRef]

5. Meng, Z.; Chen, Q.; Li, H.; See, C.H. Internal insulation condition identification for high-voltage capacitor voltage transformers based on possibilistic fuzzy clustering. *Rev. Sci. Instrum.* **2020**, *91*, 014705. [CrossRef] [PubMed]

6. Rocha, G.V.S.; de Barradas, R.P.S.; Muniz, J.R.S.; Bezerra, U.H.; de Araújo, I.M.; de da Costa, D.S.A.; da Silva, A.C.; Nunes, M.V.A.; Silva, J.S. Optimized surge arrester allocation based on genetic algorithm and ATP simulation in electric distribution systems. *Energies* **2019**, *12*, 4110. [CrossRef]

7. Mahmood, F.; Rizk, M.E.M.; Lehtonen, M. Risk-based insulation coordination studies for protection of medium-voltage overhead lines against lightning-induced overvoltages. *Electr. Eng.* **2019**, *101*, 311–320. [CrossRef]

8. Hidaka, T.; Shiota, K.; Ishimito, K.; Asakawa, A. Effect of grounding resistance connected to surge arresters in power distribution lines on lightning surge behavior observed in customer's equipment. *Electr. Eng. Jpn.* **2016**, *195*, 1–10. [CrossRef]

9. Sabiha, N.A.; Lehtonen, M. Overvoltage spikes transmitted through distribution transformers due to MV spark-gap operation. In Proceedings of the 2010 Electric Power Quality and Supply Reliability Conference, Kuressaare, Estonia, 16–18 June 2010; pp. 207–214.

10. Heine, P.; Lehtonen, M.; Oikarinen, A. Overvoltage protection, faults and voltage sags. In Proceedings of the 2004 11th International Conference on Harmonics and Quality of Power (IEEE Cat. No.04EX951), Lake Placid, NY, USA, 12–15 September 2004; pp. 100–105.

11. Borghetti, A.; Ferraz, G.M.F.; Napolitano, F.; Nucci, C.A.; Piantini, A.; Tossani, F. Lightning protection of a multi-circuit HV-MV overhead line. *Electr. Power Syst. Res.* **2020**, *180*, 106119. [CrossRef]

12. Grebovic, S.; Balota, A.; Oprasic, N. Lightning outage performance of power distribution line located in Mountain Lovćen area. In Proceedings of the 2020 9th Mediterranean Conference on Embedded Computing (MECO), Budva, Montenegro, 8–11 June 2020; pp. 1–4.

13. Mahmood, F.; Rizk, M.E.M.; Lehtonen, M. Evaluation of lightning overvoltage protection schemes for pole-mounted distribution transformers. *Int. Rev. Electr. Eng.* **2015**, *10*, 616. [CrossRef]

14. Sabiha, N.A.; Lehtonen, M. Lightning-induced overvoltages transmitted over distribution transformer with MV spark-gap operation—Part II: Mitigation using LV surge arrester. *IEEE Trans. Power Deliv.* **2010**, *25*, 2565–2573. [CrossRef]

15. Somogyi, A.; Vizi, L. Overvoltage protection of pole mounted distribution transformers. *Period. Polytech. Electr. Eng.* **1997**, *41*, 27–40.

16. ABB Switzerland Ltd. *Application Note 2.0—Metal-Oxide Surge Arresters in Medium-Voltage Systems Overvoltage Protection of Transformers*; ABB Switzerland Ltd.: Wettingen, Switzerland, 2018.

17. Bogarra, S.; Orille, L.À.; Àngela, M. Surge arrester's location using fuzzy logic techniques. In Proceedings of the 17th International Conference on Electricity Distribution, Barcelona, Spain, 12–15 May 2003; pp. 1–5.

18. de Barradas, R.P.S.; Rocha, G.V.S.; Muniz, J.R.S.; Bezerra, U.H.; Nunes, M.V.A.; Silva, J.S. Methodology for analysis of electric distribution network criticality due to direct lightning discharges. *Energies* **2020**, *13*, 1580. [CrossRef]

19. Opsahl, A.M.; Brookes, A.S.; Southgate, R.N. Lightning protection for distribution transformers. *Trans. Am. Inst. Electr. Eng.* **1932**, *51*, 245–251. [CrossRef]

20. What is the Transformer BIL (Basic Insulation Level)? Available online: https://openspark.ca/2017/06/05/what-is-the-transformer-bil-basic-insulation-level/ (accessed on 15 August 2020).

21. IEEE Std 1410-2010 (Revision of IEEE Std 1410-2004). In *IEEE Guide for Improving the Lightning Performance of Electric Power Overhead Distribution Lines*; IEEE: New York, NY, USA, 2010.

22. Mahmood, F.; Elkalashy, N.I.; Lehtonen, M. Modeling of flashover arcs in medium voltage networks due to direct lightning strikes. *Int. J. Electr. Power Energy Syst.* **2015**, *65*, 59–69. [CrossRef]

23. Ghania, S.M. Grounding systems under lightning surges with soil ionization for high voltage substations by using two layer capacitors (TLC) model. *Electr. Power Syst. Res.* **2019**, *174*, 105871. [CrossRef]

24. IEEE Std 1313.2-1999. In *IEEE Guide for the Application of Insulation Coordination*; IEEE: New York, NY, USA, 1999; pp. 1–68.

25. Resende, F.O.; Lopes, P.J.A. Using low-voltage surge protection devices for lightning protection of 15/0.4 kV pole-mounted distribution transformer. *CIRED Open Access Proc. J.* **2017**, *2017*, 888–892. [CrossRef]

26. Takahashi, A.; Hidaka, T.; Ishimoto, K.; Asakawa, A. Influence of grounding resistance connecting to surge arrester on effectiveness of lightning protection caused by direct hit for power distribution lines. *Electr. Eng. Jpn.* **2012**, *179*, 10–22. [CrossRef]

27. Paolone, M.; Nucci, C.A.; Petrache, E.; Rachidi, F. Mitigation of lightning-induced overvoltages in medium voltage distribution lines by means of periodical grounding of shielding wires and of surge arresters: Modeling and experimental validation. *IEEE Trans. Power Deliv.* **2004**, *19*, 423–431. [CrossRef]

28. Electromagnetic Transients Program- the Restructure Version (EMTP-RV). Available online: https://www.emtp-software.com/en (accessed on 15 August 2020).

29. Working Group 01 (Lightning) of Study Committee 33 (Overvoltages and Insulation Co-ordination). In *Guide to Procedures for Estimating the Lightning Performance of Transmission Lines*; Technical Brochure NO. 063: Paris, France, 1991.

30. Dudurych, I.M.; Gallagher, T.J.; Corbett, J.; Val Escudero, M. EMTP analysis of the lightning performance of a HV transmission line. *IEEE Proc. Gener. Transm. Distrib.* **2003**, *150*, 501. [CrossRef]

31. Wahlberg, M.; Rönnberg, S. Currents in power line wood poles. In Proceedings of the 21st International Conference on Electricity Distribution, Frankfurt, Germany, 6–9 June 2011; pp. 1–4.

32. Bakar, A.H.A.; Talib, D.N.A.; Mokhlis, H.; Illias, H.A. Lightning back flashover double circuit tripping pattern of 132kV lines in Malaysia. *Int. J. Electr. Power Energy Syst.* **2013**, *45*, 235–241. [CrossRef]

33. Martinez-Velasco, J.A.; González-Molina, F. Calculation of power system overvoltages. In *Transient Analysis of Power Systems*; John Wiley & Sons, Ltd.: Chichester, UK, 2014; pp. 100–194.

34. Marti, J. Accurate modelling of frequency-dependent transmission lines in electromagnetic transient simulations. *IEEE Trans. Power Appar. Syst.* **1982**, *101*, 147–157. [CrossRef]

35. Omidiora, M.A. Modeling and Experimental Investigation of Lightning Arcs and Overvoltages for Medium Voltage Distribution Lines. Ph.D. Thesis, Aalto University, Espoo, Finland, 2011.

36. Ametani, A.; Matsuoka, K.; Omura, H.; Nagai, Y. Surge voltages and currents into a customer due to nearby lightning. *Electr. Power Syst. Res.* **2009**, *79*, 428–435. [CrossRef]

37. GRASH. *Webinar on Lightning Stroke Analysis-EMTP Group (Power System Transient Analysis)*; GRASH: New Orleans, LA, USA, 2015.

38. Banjanin, M.S. Application possibilities of special lightning protection systems of overhead distribution and transmission lines. *Int. J. Electr. Power Energy Syst.* **2018**, *100*, 482–488. [CrossRef]

39. Cao, T.X.; Pham, T.; Boggs, S. Computation of tower surge impedance in transmission line. In Proceedings of the 2013 IEEE Electrical Insulation Conference (EIC), Ottawa, ON, Canada, 2–5 June 2013; pp. 77–80.

40. Banjanin, M.S.; Savić, M.S. Some aspects of overhead transmission lines lightning performance estimation in engineering practice. *Int. Trans. Electr. Energy Syst.* **2016**, *26*, 79–93. [CrossRef]

41. Savadamuthu, U.; Udayakumar, K.; Jayashankar, V. Modified disruptive effect method as a measure of insulation strength for non-standard lightning waveforms. *IEEE Trans. Power Deliv.* **2002**, *17*, 510–515. [CrossRef]

42. IEEE Working Group 3.4.11; Application of Surge Protective Devices Subcommittee- Surge Protective Device Committee. Modeling of metal oxide surge arresters. *IEEE Trans. Power Deliv.* **1992**, *7*, 302–309.

43. Christodoulou, C.A.; Assimakopoulou, F.A.; Gonos, I.F.; Stathopulos, I.A. Simulation of metal oxide surge arresters behavior. In Proceedings of the 2008 IEEE Power Electronics Specialists Conference, Rhodes, Greece, 15–19 June 2008; pp. 1862–1866.

44. Cigre Working Group 02; (Internal Overvoltages) of Study Committee 33. *Guidelines for Representation of Network Elements When Calculating Transients*; (Internal Overvoltages) of Study Committee 33: Paris, France, 1990; 30p.

45. Pourakbari-Kasmaei, M.; Mahmood, F.; Krbal, M.; Pelikan, L.; Orságová, J.; Toman, P.; Lehtonen, M. Evaluation of filtered spark gap on the lightning protection of distribution transformers: Experimental and simulation study. *Energies* **2020**, *13*, 3799. [CrossRef]

46. Paulino, J.O.S.; Barbosa, C.F.; Lopes, I.J.S.; do Boaventura, W.C.; de Miranda, G.C. Indirect lightning performance of aerial distribution lines considering the induced-voltage waveform. *IEEE Trans. Electromagn. Compat.* **2015**, *57*, 1123–1131. [CrossRef]

47. Darveniza, M. The generalized integration method for predicting impulse volt-time characteristics for non-standard wave shapes-a theoretical basis. *IEEE Trans. Electr. Insul.* **1988**, *23*, 373–381. [CrossRef]

48. Pourakbari-Kasmaei, M.; Mahmoud, F.; Krbal, M.; Pelikan, L.; Orságová, J.; Toman, P.; Lehtonen, M. Transformer Winding Capacitance Measurement. Available online: https://drive.google.com/file/d/1JbpTBkmuW2-AxfnBX5mN0JIF9Z_iLaN-/view?usp=sharing (accessed on 15 August 2020).

49. Pourakbari-Kasmaei, M.; Mahmoud, F.; Krbal, M.; Pelikan, L.; Orságová, J.; Toman, P.; Lehtonen, M. Transformer with Serial Inductors and Protection Spark Gap. Available online: https://drive.google.com/file/d/1ztdvBW2QNNZx1t6XsAY6IMTzq-nxAlVc/view?usp=sharing (accessed on 15 August 2020).

50. Miyazaki, S.; Goshima, H.; Amano, T.; Shinkai, H.; Yashima, M.; Wakimoto, T.; Ishii, M. Uncertainty of peak-value measurement of lightning impulse high voltage by national-standard-class measuring system. In Proceedings of the XVII International Symposium on High Voltage Engineering, Hannover, Germany, 22–26 August 2011; pp. 1–6.

51. IEEE Std 1313-1993. In *IEEE Standard for Power Systems—Insulation Coordination*; IEEE: New York, NY, USA, 1993.

52. Technical Specification of Transformer. Available online: https://procurement-notices.undp.org/view_file.cfm?doc_id=104038 (accessed on 15 August 2020).

53. Phoenix Contact. The Basics of Surge Protection: From the Generation of Surge Voltages Right through to a Comprehensive Protection Concept. Available online: http://media.automation24.com/manual/nl/5131327_TT_Basics_Surge_protection_EN.pdf (accessed on 15 August 2020).

Article

Resonant Power Frequency Converter and Application in High-Voltage and Partial Discharge Test of a Voltage Transformer

Banyat Leelachariyakul and Peerawut Yutthagowith *

King Mongkut's Institute of Technology Ladkrabang, School of Engineering, 1 Chalongkrung Rd., Ladkrabang, Bangkok 10520, Thailand; klbanyat@gmail.com
* Correspondence: peerawut.yu@kmitl.ac.th; Tel.: +66-(0)2-329-8330

Abstract: This paper presents application of a resonant power frequency converter for high-voltage (HV) and partial discharge (PD) test of a voltage transformer. The rating voltage, power, and frequency of the system are 70 kV$_{rms}$, 40 kVA, and 200 Hz, respectively. The testing system utilized the converter feeding to an HV testing transformer connected to a conventional partial discharge detection system. The converter system comprising a rectifier and insulated-gate bipolar (IGBT) switches with the H-bridge configuration was applied as a low-voltage source instead of a conventional motor-generator test set which requires large space and high cost. The requirements of the test according to the standards are quality of the test voltage and the background noise level. The required voltage must have the different voltage (DV) and total harmonic distortion (THD_v) in the acceptable values of less than 5%. The DV is defined as the difference of the root mean square and peak voltages in percent. The required background noise level must be lower than 2.5 pC. Simulations and experiments were performed for verification of the developed system performance in comparison with those of the previously developed system based on the pulse width modulation converter. It is found that the developed system can provide the testing voltage with the DV and the THD_v of lower than 1% and the background noise level of lower than 1 pC. Considering this achievement of promising performance, the developed system is an attractive choice for the HV and PD testing of voltage transformers in real practice.

Keywords: background noise; high-voltage and partial discharge test; resonant power frequency converter; switching interference; voltage transformer

Citation: Leelachariyakul, B.; Yutthagowith, P. Resonant Power Frequency Converter and Application in High-Voltage and Partial Discharge Test of a Voltage Transformer. *Energies* 2021, *14*, 2014. https://doi.org/10.3390/en14072014

Academic Editor: Dan Doru Micu

Received: 18 February 2021
Accepted: 22 March 2021
Published: 5 April 2021

Publisher's Note: MDPI stays neutral with regard to jurisdictional claims in published maps and institutional affiliations.

1. Introduction

The main problem of the high-voltage (HV) and partial discharge (PD) test of voltage transformers is the applied voltage, which is higher than the transformer rating voltage. If such a voltage level with the rating frequency (50 Hz or 60 Hz) is applied to the test, it will lead to core saturation, high current consumption, and applied testing voltage distortion. For avoiding such conditions, the international standard defines the quality of the testing voltage with the different voltage (the root mean square value and peak value divided by square root of two) and total harmonic distortion, which are not higher than 5%. Under the test conditions, the applied voltage in such tests is not higher than two times the rating voltage of the transformer, so the applied voltage should have a frequency higher than two times that of the rating frequency. For safety reasons, the frequency used in the test is from 100 to 400 Hz. It is found that the frequency of 200 Hz is sufficient in all tests of the voltage transformers. Therefore, the 200 Hz frequency was utilized in all considered simulations and experiments. In the past, the motor and generator test set was applied in the test. However, the test set is quite costly, and it requires a large space for installation.

The high-voltage (HV) test is a crucial issue for the verification of the performance of the HV equipment in design and construction processes. Examples of such tests are insulation voltage withstand test, power loss measurement, dielectric loss measurement, and partial discharge measurement.

According to IEC 60270:2015, [1] partial discharges (PDs) are defined as localized electrical discharges that only partially bridge the insulation between conductors, and they are caused by local electrical stress concentrations in the insulation or on the surface of the insulation. Generally, such discharges appear as pulses with durations of less than 1 µs. PD measurement is the most important test used in the evaluation of insulation performance and life. The PDs are classified into three main types. The first type of the PD, named corona discharge, can occur in areas with sharp edges and high electric field stress. The second type originates from defects of the internal insulation material, such as bubbles and voids. If these defects have lower insulation levels than the main insulation material, under sufficient electrical field stress, PD can occur. The last PD type, called surface discharge, occurs at the boundary between two materials.

As the simplest way to generate HV for HV testing, [2] a voltage regulator connected with a power source from a low-voltage grid or a generator is applied to an HV testing transformer at the low-voltage side. The HV from the transformer is applied to a test object. Due to the nonlinear characteristic of the electrical and electronic equipment, the disturbance is generated in the electrical system, and it sometimes causes power quality problems, i.e., the voltage distortion and the background noise level. According to IEC 60060-1, [3] the testing voltage should be an almost purely sinusoidal waveform of which the total harmonic distortion voltage (THD_v), defined by Equation (1), and different voltage (DV; peak voltage divided by $\sqrt{2}$ and RMS voltage) should be less than 5%.

$$THD_v = \frac{1}{V_{p1}}\sqrt{\sum_{i=2}^{50} V_{pi}^2} \tag{1}$$

Here, V_{p1} is the peak voltage of the fundamental frequency and V_{pi} is the peak voltage of the *i*th harmonic frequency. Figure 1 shows the test voltage applied to the voltage transformer (VT) and PD pattern during the HV and PD test. It is found that the DV and THD_v are higher than 5%. Additionally, the background noise level is higher than 2.5 pC (the acceptable background noise for the HV and PD test of the voltage transformers) [4,5]. To avoid the voltage distortion, and the unacceptable background noise level and to satisfy the standard requirements, an additional measure such as a voltage filter is necessary to be applied to the system.

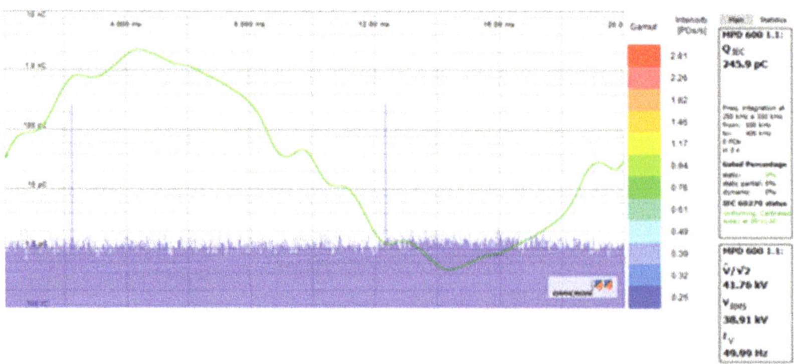

Figure 1. Test voltage waveform and the partial discharge (PD) pattern during the high-voltage (HV) test.

For the HV test of voltage transformers (VTs) and to confirm the insulation performance of the test object, it is necessary to raise the voltage to be higher than the rating voltage. If the applied testing voltage with the rated frequency is used in the test, the core saturation of the voltage transformer can occur and the applied voltage may be distorted. For a better understanding, an example case should be considered. For the partial discharge test of a VT with the rating voltage (U_r) of 24 kV, the pre-stress voltage condition with a withstand voltage (U_t) of 80% is applied to the VT, and then the testing voltage is decreased to the level of $1.2U_r$ (28.8 kV) to record the PD activity. The procedure of applying voltage in the PD test is shown in Figure 2. Therefore, saturation of the iron core of VT is avoided by applying voltage with a frequency higher than the rated frequency of the VT. If the voltage with a power frequency of 50 Hz is applied to the HV side of the VT, the saturation effect of the VT will influence the distortion of the applied voltage waveform, as shown in Figure 1, and the DV and THD_v will be higher than 5%. Aside from the nonlinear voltage, the core saturation can cause thermal runaway and explosion of PT. Therefore, to avoid the core saturation effect, a test voltage with a frequency higher than twice the power frequency is necessary in the test; the frequency of 200 Hz is a good candidate for all VTs with the rating frequencies of 50 and 60 Hz.

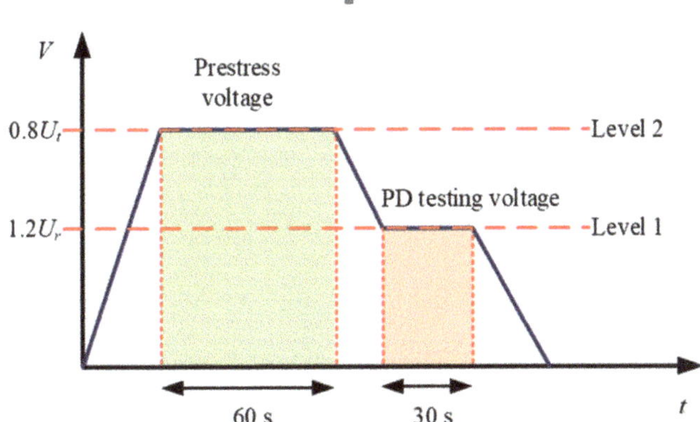

Figure 2. Procedure of voltage application for the PD test of voltage transformers (VTs).

Nowadays, power electronics technology has been applied in many practical fields, such as home appliances, automotives and traction, renewable energy, and HV transmission and distribution systems. In HV generation for testing, the power electronics converter is a powerful and efficient tool for AC/DC, DC/AC, and AC/AC power conversion. However, the crucial problem of the application of the power converter in the PD test is the interference signal originating from fast switching of power electronic devices. It causes the PD detection system to have background noise much higher than the acceptable level. For example, in the PD tests of VTs insulated with oil, the acceptable PD level is only 5 pC [4,5]. Consequently, the testing system including the PD detection system should have a background noise level below 50% of the acceptable PD level (2.5 pC). Most commercial power converters have an additional filter for eliminating undesired harmonic voltage and obtaining a voltage waveform close to a pure sinusoidal wave. However, the filter cannot eliminate the interference in the PD measurement completely, leading the background noise in the PD test to be higher than the acceptable level.

Recently, there are attempts for development of the HV and PD test of voltage transformer. In [6–9], the power frequency converter based on the pulse width modulation (PWM) techniques was developed as the low-voltage source applied to the HV testing transformer for the HV and PD test. In [10,11], the analysis of effect of the additional filter for the reduction in the background noise in the HV and PD test was presented. In [12], the analysis of the PWM converter with the additional filter was presented, and the best switching frequency in terms of the quality of the generated voltage and background noise level were investigated. However, it is found that the developed system in [12] required the additional filter with quite large capacitance and high power consumption.

For overcoming such problems, the HV and PD testing system based on a resonant power frequency converter is proposed in this paper. Simulations were utilized for the design of the system, and experiments were performed to verify the validity of the developed system. The system performances in terms of the testing voltage quality and background noise level were investigated and analyzed in comparison with those of the previous developed system based on the pulse width modulation converter. It is found that the developed system performances are much superior to those of the previous system. The different voltage and total harmonic distortion of the testing voltage are less than 1%, and the background noise level is less than 1 pC at the rating voltage. From this achievement in terms of the system performances, the developed system is an attractive choice for the HV and PD testing of voltage transformers in the real practice.

2. Development of HV Testing System for PD Tests

The developed system with the equivalent circuit shown in Figure 3 is composed of a power frequency converter, an additional inductor, an additional capacitor, an HV testing transformer, and a partial discharge detection system. Z_f, C_k, CD, CC, and MI stand for the internal impedance of the testing transformer, the coupling capacitor, the coupling device or the measuring impedance, the measuring cable, and the measuring instrument, respectively.

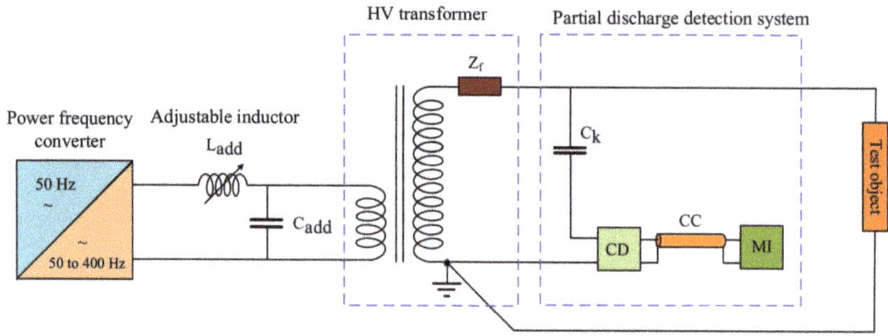

Figure 3. Partial discharge testing system.

2.1. Power Frequency Converter

The power frequency converter based on the H-bridge configuration, as shown in Figure 4, was developed. The converter is composed of a rectifier and H-bridge insulated-gate bipolar transistors (IGBTs), which can be controlled to generate either a square wave or pulse width modulation (PWM) waveforms.

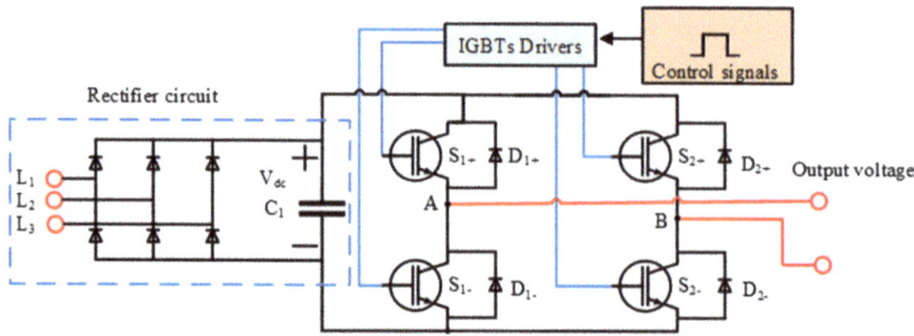

Figure 4. High-voltage source power converter generator.

The unipolar PWM technique [13–15] is based on operation of four control switches (S1, S2, S3, and S4) and four diodes (D1, D2, D3, and D4) as shown in Table 1. The AC output voltage waveform can instantaneously take one of the following three voltage levels: $+V_{dc}$, $-V_{dc}$, or 0.

Table 1. Switch states in full-bridge single-phase voltage source inverter unipolar pulse width modulation (PWM).

State	Switch Conduction Status		Conduction Status of IGBT and Diode		V_o
	ON	OFF	$I_o > 0$	$I_o < 0$	
1	S_{1+}, S_{2-}	S_{1-}, S_{2+}	S_{1+}, S_{2-}	D_{1+}, D_{2-}	$+V_{dc}$
2	S_{1+}, S_{2+}	S_{1-}, S_{2-}	S_{1+}, D_{2+}	D_{1+}, S_{2+}	0
3	S_{1-}, S_{2+}	S_{1+}, S_{2-}	D_{1-}, D_{2+}	S_{1-}, S_{2+}	$-V_{dc}$
4	S_{1-}, S_{2-}	S_{1+}, S_{2+}	D_{1-}, S_{2-}	S_{1-}, D_{2-}	0

The controlled and AC reference voltage waveforms are shown in Figure 5a, and the output signal is shown in Figure 5b.

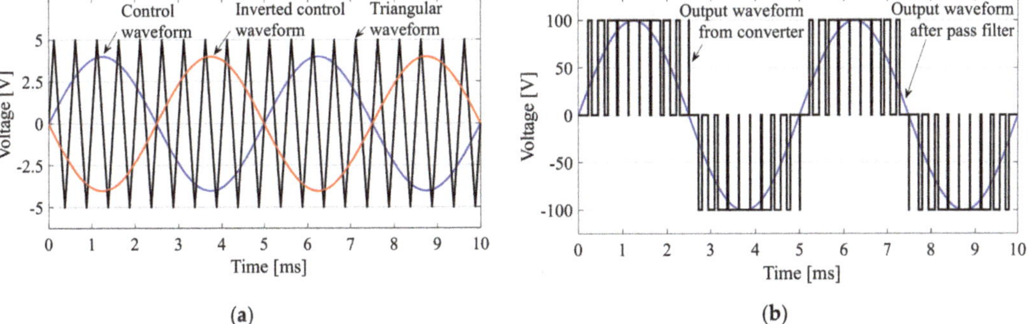

Figure 5. Generated PWM waveform from the PWM converter. (a) Pulse width modulation in the unipolar converter. (b) Output waveform of the power converter.

For generation of the PWM waveform, the triangular waveform is set to be from 1.2 to 20 kHz. However, from the study in [12], it was found that the best performance of the PWM converter occurred at the switching frequency of 3.2 kHz. For generation of the square wave used in the proposed resonant circuit, the frequency of the triangular waveform is set to be the same as that of the control waveform.

2.2. Partial Discharge Detection System

The developed partial discharge detection system is composed of a coupling capacitor (C_k) connected in series with the measuring impedance (coupling device). The capacitance C_k was selected to be 1 nF, and the measuring impedance was designed to have the band-pass characteristic with the equivalent circuit shown in Figure 6. The transfer impedance characteristic with the low and high cut-off frequencies of 30 kHz and 20 MHz, as shown in Figure 7. For avoiding the undesired noise signal in the HV testing environment, the standard [1] recommends using the band pass filter for a quasi-integration of the charge determination, and the acceptable range of cut-off frequencies of the filter are also provided [1]. In this paper, the digital band pass filter with the cut-off frequencies of 100 kHz and 400 kHz is utilized according to the standard requirement. The transfer function of the filter and the transfer impedance with the considered filter are also presented in Figure 7.

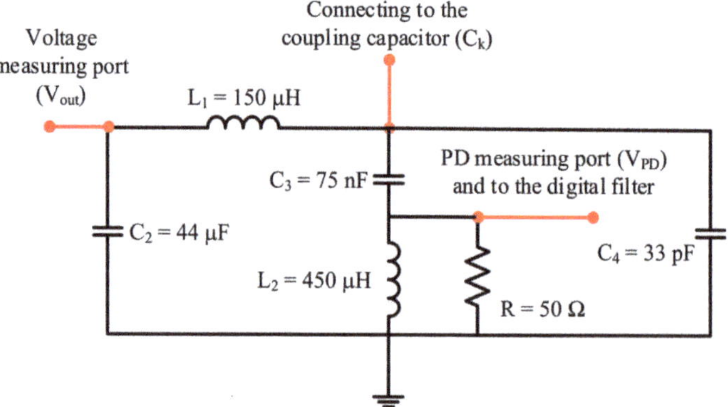

Figure 6. Equivalent circuit of the measuring impedance or the coupling device (CD).

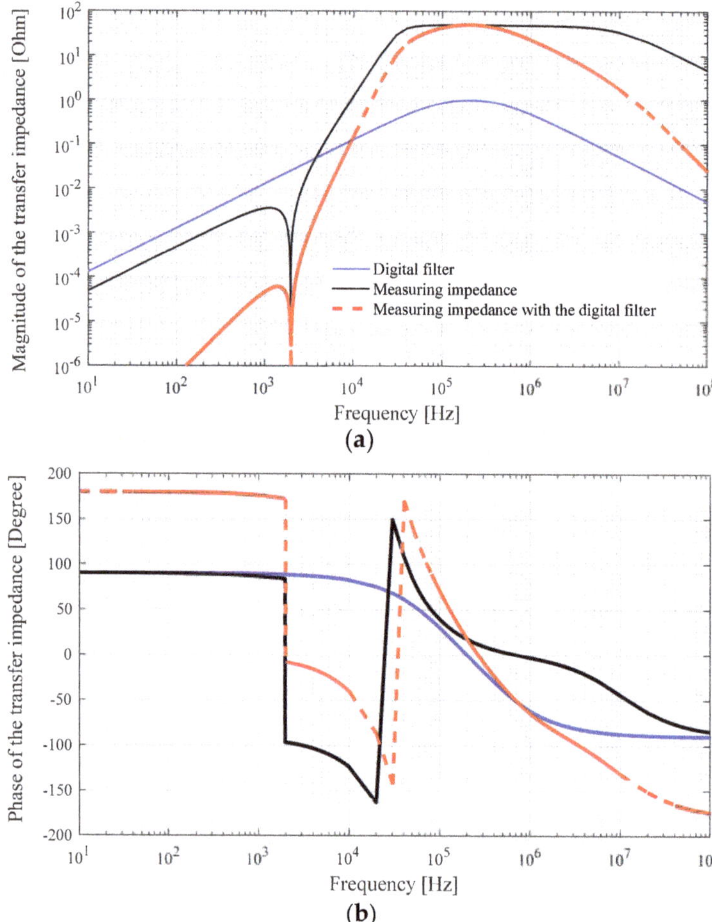

Figure 7. Transfer impedance of the measuring impedance in the frequency domain. (**a**) Magnitude (**b**) Phase in degree.

The equivalent circuit of the PD detection system with the HV testing transformer is shown in Figure 8. The crucial circuit parameters of the testing system are expressed in Table 2. The transfer function of the output voltage (V_{out}/V_{in}) can be calculated as shown in Figure 9 which shows that this transfer function has the characteristic of a low-pass filter. As shown in Figure 9, the cut-off frequency is about 1.2 kHz. It is noted that the total impedance of the HV testing transformer with the rating voltage of 460 V/75 kV and the rating power of 40 kVA is transferred to the HV side of the transformer. The circuit parameters are expressed in Figure 8, and the voltage transformer under test [16] or the test object can be represented well with a high impedance, which has an effect on the test circuit. When the 1 pC calibrator pulse current is injected into the system in the calibration process, the peak voltage response at the PD port and filtered peak voltage by the digital band pass filter are about 2.0 mV and 0.115 mV, respectively. The results of the response voltages are shown in Figure 10.

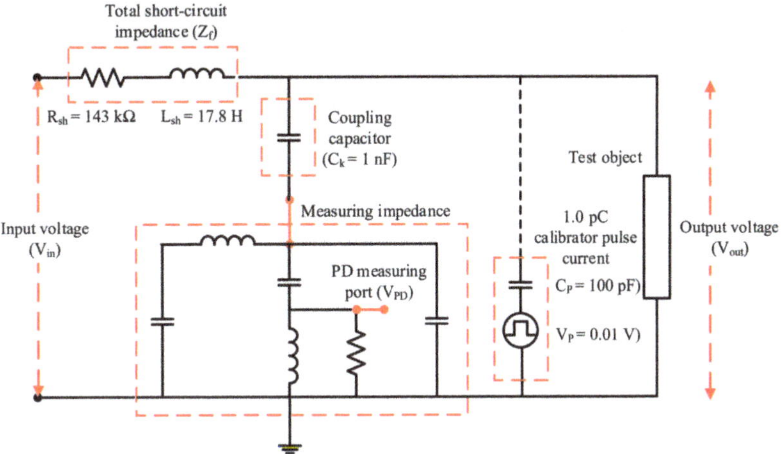

Figure 8. Simplified equivalent circuit of the HV and PD tests.

Table 2. Circuit parameters of the testing system.

Circuit Parameters	Transfer to the LV Side	Transfer to the HV Side
Testing transformer series resistance (R_{sh})	5.38 Ω	143 kΩ
Testing transformer series inductance (L_{sh})	0.67 mH	17.8 H
Coupling capacitance (C_k)	26.8 µF	1 nF

For a better understanding of the problem of using the PWM frequency converter in the HV and PD test, we should consider the case in which the 200 Hz PWM voltage of 100 kV with the switching frequency of 3.2 kHz is supplied to the system without an additional filter as shown in Figure 8. The voltage across the test object has small oscillation, as shown in Figure 11, and its THD_v is 0.93%. The PWM voltage also affects the voltage at the PD measuring port. The peak voltages at the PD port and the filtered one are of 570 mV and 370 mV, respectively. As shown in Figure 12, the interference voltages are much higher than the peak voltage response of a 1 pC calibrator pulse. It can be seen that the band pass filter is not very effective for mitigation of the interference signal from the switching voltage. Therefore, the mean value for mitigating the interference from the PWM frequency converter is required for the PD test in real practice. In this paper, the additional inductor and capacitor are proposed to connect at the LV side of the HV testing transformer for obtaining the resonant condition. The analysis of the additional circuit components will be presented in the next Section 2.3.

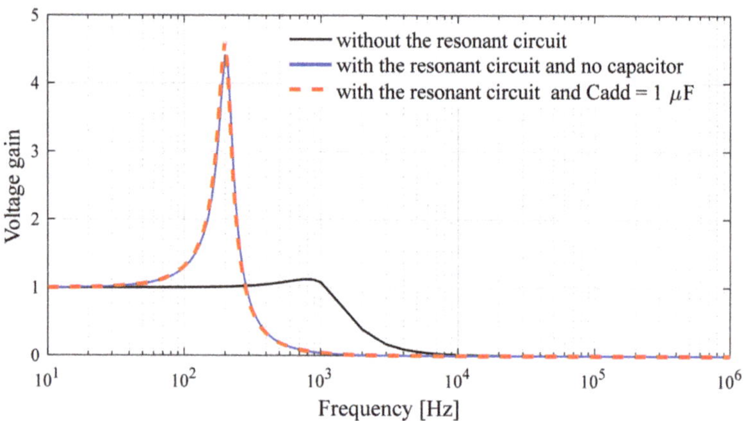

Figure 9. The calculated and normalized transfer function of the testing voltage with and without additional resonant circuit.

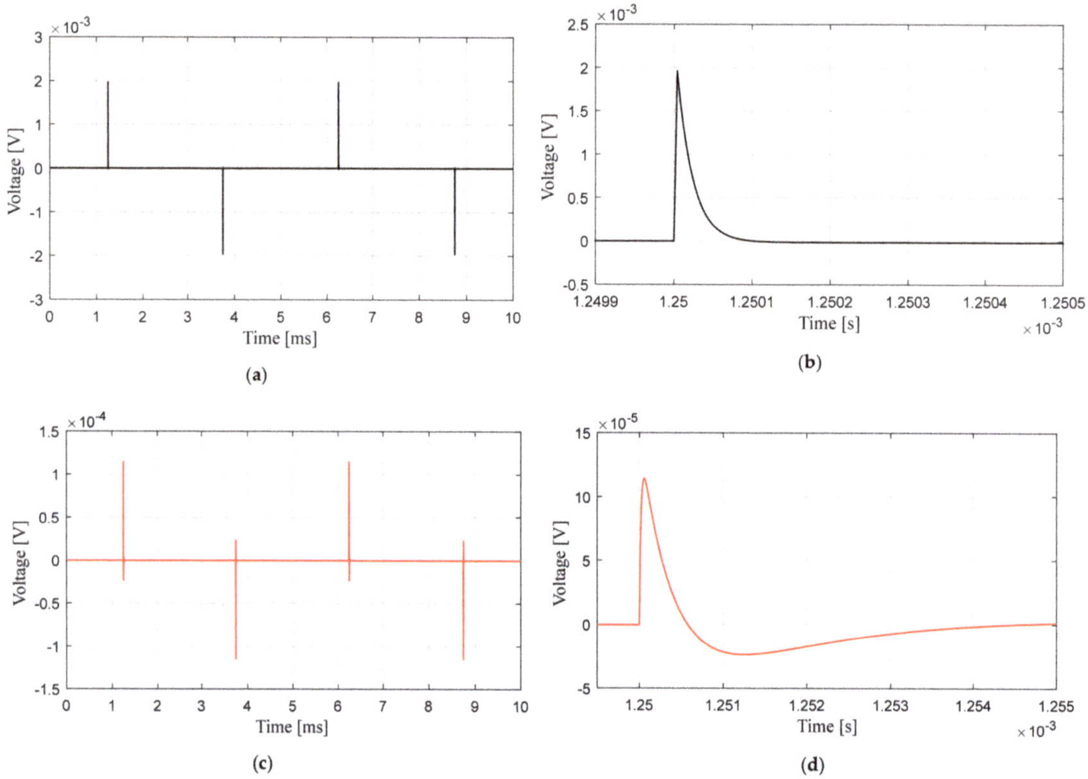

Figure 10. Voltage response of 1 pC calibrator pulse current (**a**) at the PD port with long span time, (**b**) at the PD port with short span time, (**c**) filtered by the band pass filter with long span time, (**d**) filtered by the band pass filter with short span time.

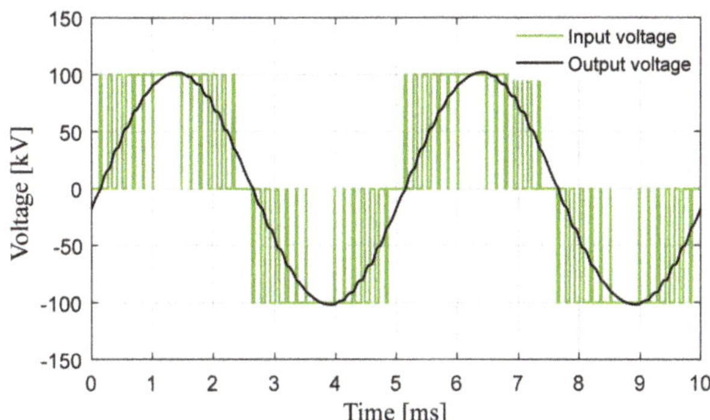

Figure 11. Input and output voltage waveforms.

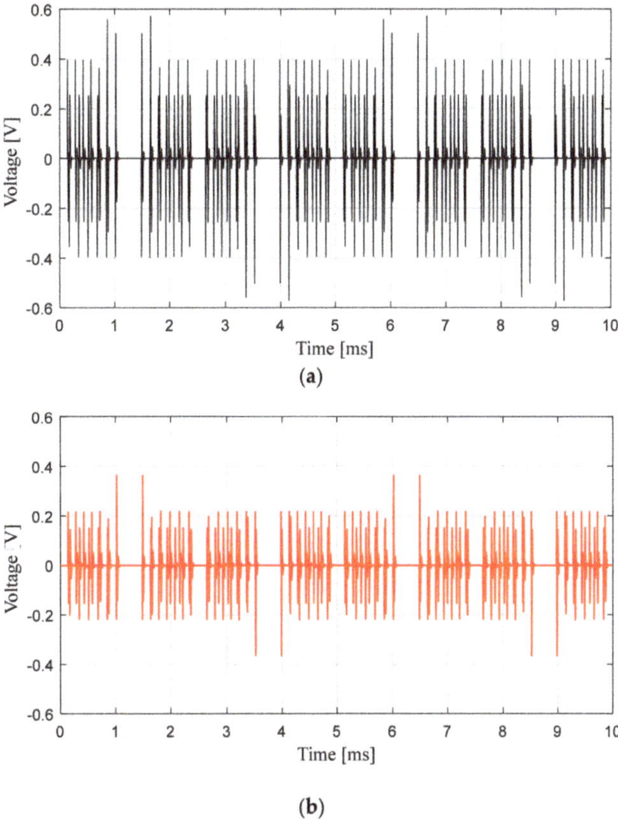

Figure 12. Voltage response at the PD port in the case of applied PWM voltage at 100 kV$_\text{p}$. (**a**) At the PD port (**b**) filtered by the band pass filter.

2.3. Additional Circuit for Circuit Resonance

To reduce the noise level, the additional filter is required as in the previous studies [6–12]. In this paper, an additional adjustable inductor and capacitor are proposed to add in the low-voltage side of the testing transformer for obtaining the resonant condition of the test circuit as shown in Figure 3. The inductance was adjusted to obtain the resonant condition at the frequency of 200 Hz. With the circuit parameters of the testing system, the series resonant condition is occurred by Equation (2), where L_t is total inductance and C_t is total capacitance. In the considered system, the total capacitance can be approximated as the coupling capacitance (C_k). The required additional inductance (L_{add}) can be calculated by Equations (2) and (3), and the result is about 23 mH.

$$L_t = \frac{1}{\omega^2 C_t} \tag{2}$$

$$L_t = L_{add} + L_{sh} \tag{3}$$

The voltage gain at the resonant condition is equivalent to the quality factor (Q) as expressed in Equation (4), where R_t is total series resistance composed of internal resistance (R_{int}) of the additional inductor and series resistance of the testing transformer (R_{sh}).

$$Q = \frac{\sqrt{L_t/C_t}}{R_t} \tag{4}$$

In this paper, the required Q was set to be not less than 4 which is sufficient for obtaining the almost pure sinusoidal output voltage. It is noted that the quality factor of 4 will be confirmed to be sufficient by the simulation result in the Section 3. For the designed inductor, the internal resistance is necessary to be controlled for obtaining the desired Q. The maximum internal resistance (R_{int}) can be calculated by Equation (5), and the result is 2.0 Ω.

$$R_t = R_{int} + R_{sh} \tag{5}$$

From the calculation above, the adjustable inductor was designed by a manufacturer. The inductance of the developed adjustable inductor can be varied from 10 to 40 mH, and the internal resistance is 1.3 Ω.

The equivalent circuit of the HV and PD testing system is expressed in Figure 13. All circuit parameters in Figure 13 were transferred to the HV side of the testing transformer. It is noticed on the equivalent circuit in Figure 8 that the inductor will not affect the PD measurement if the PD occurs on the HV side. The transfer function in the form of the attenuation factor (V_{PD}/V_{in}) can be determined as the results in Figure 14a and the attenuation factor with the band pass filter recommended by the standard are expressed in Figure 14b. The input square wave voltage generated by the converter of 17.1 kV (corresponding to 100 kV output voltage) and the output voltage are expressed in Figure 15. The black line stands for the input voltage of the HV testing transformer (transferred to the HV side of the testing transformer), and the red line stands for the output voltage of the testing transformer or the voltage across the test object. It is noticed that the output voltage waveform is almost pure sine due to the occurrence of resonant condition at the frequency of 200 Hz. It is also confirmed that the quality factor of 4 is sufficient to obtain the waveform as the standard requirement.

In the results shown in Figures 16 and 17. The black line is denoted for the voltage response of 1 pC calibrator pulse current, and the red line stands for the interference signal at the output voltage of 100 kV. Without the additional capacitor (C_{add}), the interference from the square wave as the input voltage of the testing transformer is still high. Additionally, the band pass filter recommended by the standard is not particularly effective for mitigation of the attenuation of the interference signal, as shown in Figure 16. With connection of the additional capacitor (C_{add}), the interference is attenuated to be lower than the voltage response of the 1 pC calibrator pulse current. Additionally, the band pass filter recommended by the standard is quite effective for mitigation of attenuation of interference signal as it is shown in Figure 17. In Figure 17b, the maximum interference signal is 0.07 mV, which is equivalent to background noise level of 0.6 pC (0.115 mV is equivalent to the PD level of 1 pC).

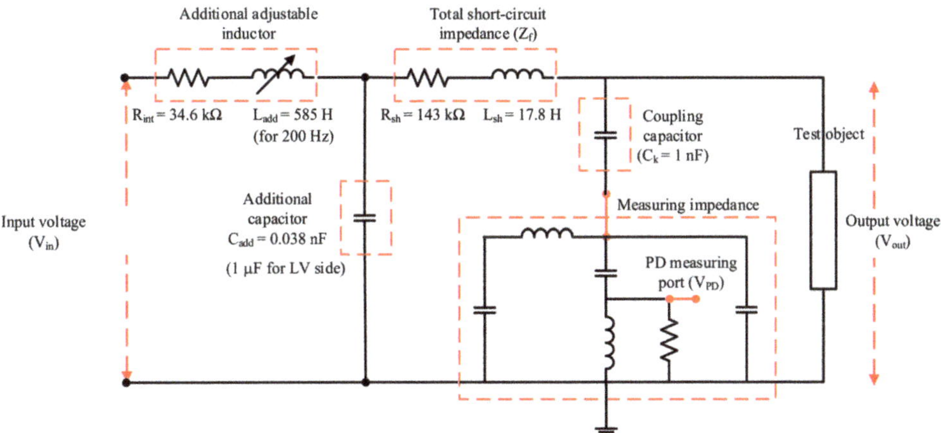

Figure 13. Equivalent circuit of the PD testing system with the resonant circuit.

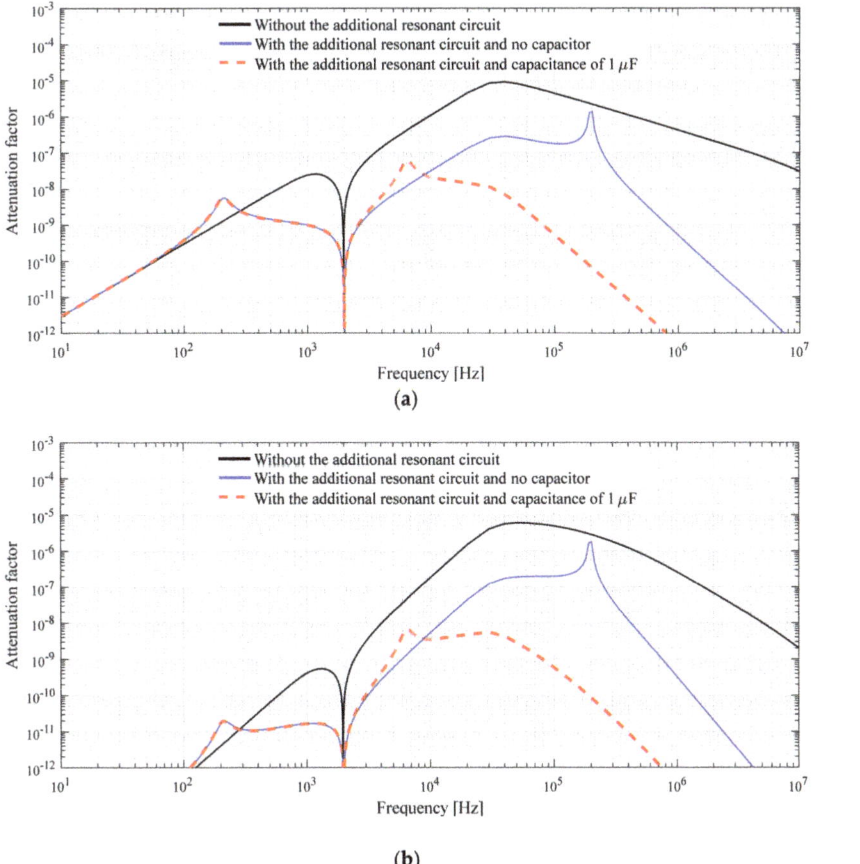

Figure 14. Attenuation factor of the HV and PD testing system. (**a**) At the PD port without the band pass filter recommended by the standard; (**b**) at PD port with the band pass filter recommended by the standard.

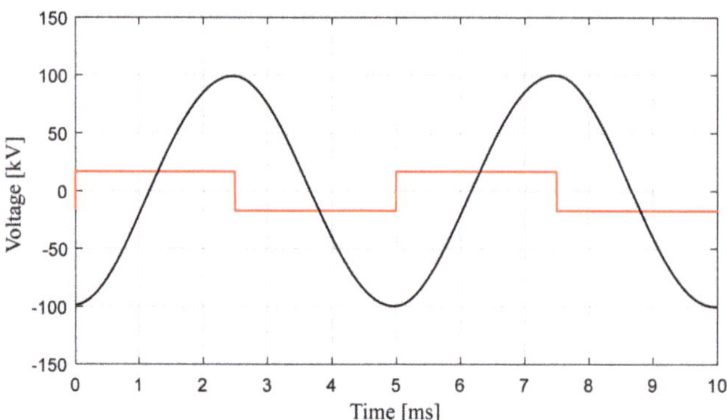

Figure 15. The input and output voltages of the testing transformer (transferred to the HV side).

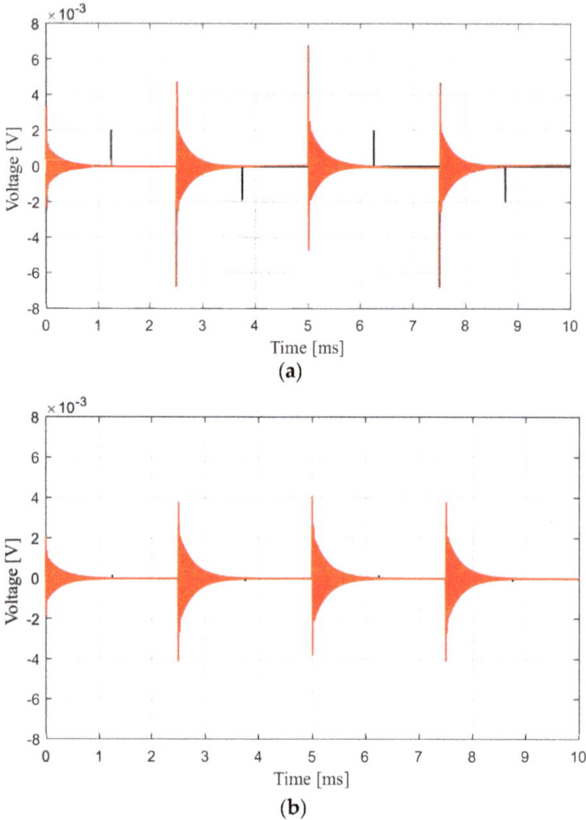

(a)

(b)

Figure 16. Comparison of the voltage responses from 1 pC calibrator pulse current and the interference signal of the resonant circuit in the case of the output voltage of 100 kV and without the additional capacitor. (**a**) At the PD port without the band pass filter recommended by the standard; (**b**) at PD port with the band pass filter recommended by the standard.

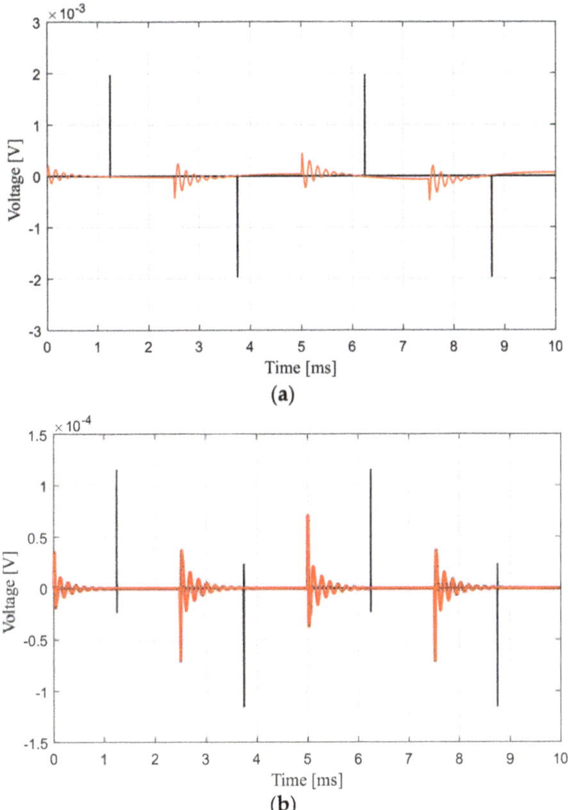

Figure 17. Comparison of the voltage responses from 1 pC calibrator pulse current and the interference signal of the resonant circuit in the case of the output voltage of 100 kV and with the additional capacitor connection. (**a**) At the PD port without the band pass filter recommended by the standard; (**b**) at PD port with the band pass filter recommended by the standard.

3. Experiments

Some experiments were carried out to investigate the performance of the developed system in the partial discharge tests. The experimental setup is shown in Figure 18. The circuit parameters were set to be the same as in Figure 13. The performances of the developed system were investigated in terms of the different voltage, total harmonic distortion of the output voltage, and the background noise.

To avoid interference signal from electromagnetic coupling in the HV laboratory, the commercial EO/OE converters (Omicron) [17] with fiber optic cables were applied with the developed system. In the real PD test, the standard PD current with the charge of 5 pC was utilized in the calibration process. The background noise without the application of the developed inverter is about 0.8 pC. The experiments are separated into two topics as follows.

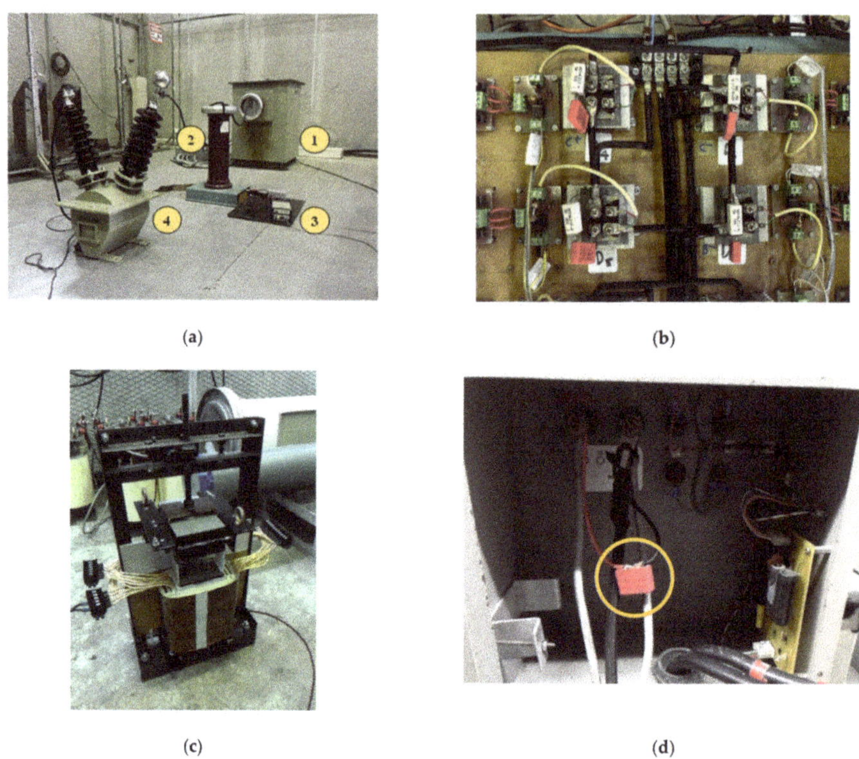

Figure 18. Experimental setup of the PD test with the VT. (**a**) Equipment in the HV testing room: (1) HV testing transformer (2) Coupling capacitor (3) Coupling device for PD detection (4) VT under test. (**b**) The developed converter, (**c**) adjustable inductor and (**d**) the additional capacitor connected at the input of the HV testing transformer.

3.1. Experiments without Test Object

The experiments in this part are performed to examine the performance of the developed system in comparison with the previous developed system (the PWM converter with the switching frequency of 3.2 kHz) [12]. There is no test object connected to the test system. As shown in Figure 19a, the PWM system without an additional filter provides very high interference. However, when the additional filter was connected to the system, at the pre-stress voltage of 40 kV$_{rms}$, the background noise was reduced significantly to be about 2.0 pC, as shown in Figure 19b. The DC input voltage supplied to the converter was about 370 V. The different voltage and THD_v are 0.66% and 0.80%, respectively.

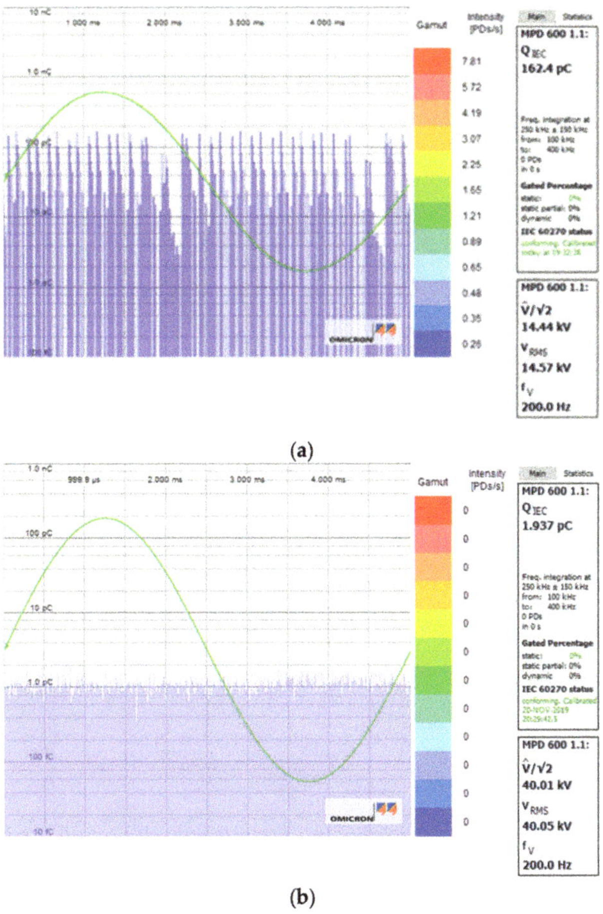

(a)

(b)

Figure 19. Experimental test results at the testing voltage of 40 kV$_{rms}$. (**a**) Without the additional filter; (**b**) with the additional filter.

When the developed resonant system was considered, in the case of no connection of the additional capacitor, the acceptable background noise (2.5 pC) was reached at the output voltage of about 5 kV, as shown in Figure 20a. This agreed with the simulation result in Figure 16b. The interference signal of 4 mV is equivalent to the background noise level of 34.78 pC (4/0.115). With the same proportion of the simulation result, the background noise level should be 2.48 pC (34.78 × 5/70). In the case of connection of the additional capacitor, the background noise level at the output voltage is still the same as that of no application of the converter to the system, as shown in Figure 20b. At the output voltage of 40 kV$_{rms}$, the DC input voltage supplied to the converter was about 83 V, which corresponds to the voltage gain of 4.5—this agreed with the simulation results. The different voltage and THD_v are 0.61% and 0.35%, respectively.

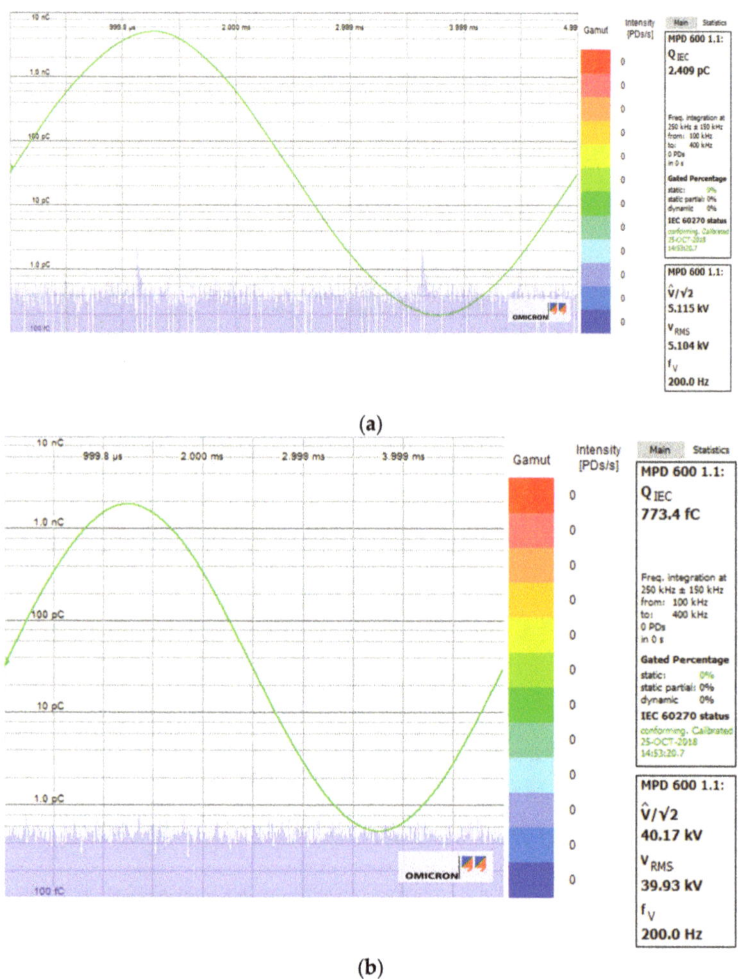

(a)

(b)

Figure 20. Experimental test results at the testing voltage of 40 kV$_{rms}$. (**a**) Without the additional capacitor; (**b**) with the additional capacitor.

3.2. Experiment with a Test Object

For confirming validity of the developed system, two voltage transformers (test object 1 and 2) were tested. The frequency of the input voltage varied from 50 to 400 Hz for examining the voltage gain. The experimental results in comparison with the simulation result without test object are expressed in Figure 21. It is found that the test object does not affect much the voltage gain. The voltage gain at the resonant frequency is still higher than in all considered cases.

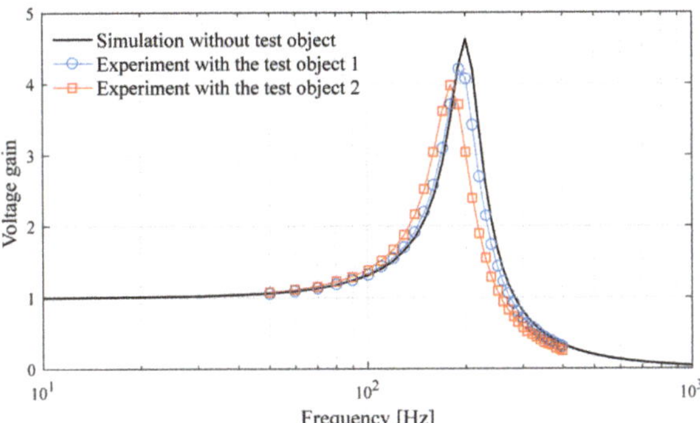

Figure 21. Comparison of the simulation and experimental voltage gains.

The other experiment was performed to confirm the performance of the developed system in the PD test with a VT. The voltage was raised up to 40 kV$_{rms}$ for the prestress condition and decreased to 30 kV$_{rms}$ for recording PD activity. The standard PD pulse with a charge of 5 pC is used for calibration in the PD test. The experimental results at the testing voltage of about 30 kV$_{rms}$ are shown in Figure 22. The different voltage (peak voltage/$\sqrt{2}$ = 28.98 kV and RMS voltage 28.79 kV) was 0.65%, the THD_v was 0.35%, and the background noise was less than 1 pC.

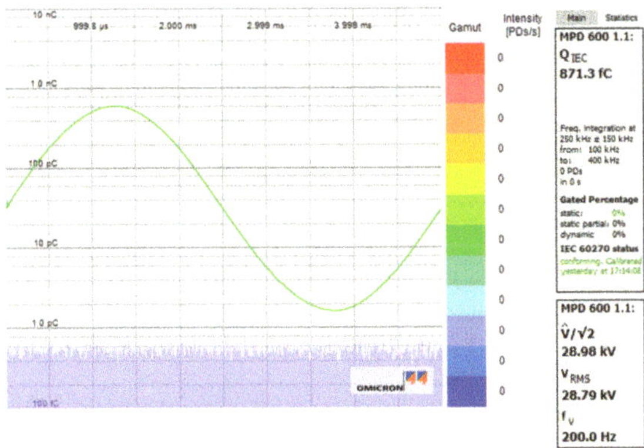

Figure 22. Experimental results with VT connected to the developed system.

4. Conclusions

The developed partial discharge testing system based on the resonant frequency converter has been presented. The performance of the developed high-voltage and partial discharge test in terms of the different voltage (DV), THD_v, and background noise in the high-voltage and partial discharge test has been investigated in comparison with those of the previously developed system based on the PWM technique. According to the simulation and experimental results, the DV was less than 1%, THD_v was less than 0.5%, and background noise was less than 1 pC. In addition, the developed system was successfully used for the real PD test on a voltage transformer with the rating voltage of 24 kV. The developed system is an attractive choice for PD tests because of its high performance and low cost compared with the motor-generator system and can be developed into a mobile system for partial discharge tests in on-site inspections. However, the limitation of the proposed system is that the proposed system is effective for the small load test object. The additional filter is still required for the test with the large load test object.

Author Contributions: Conceptualization, B.L. and P.Y.; methodology, B.L. and P.Y.; validation, B.L.; formal analysis, P.Y.; investigation, P.Y. and B.L.; writing—original draft preparation, B.L. and P.Y.; writing—review and editing, P.Y. and B.L.; supervision, P.Y. All authors have read and agreed to the published version of the manuscript.

Funding: This work was financially supported by Research and Researchers for Industries (RRI), the Thailand Research Fund (TRF) No. PHD60I0023.

Acknowledgments: This work was financially supported by Research and Researchers for Industries (RRI), the Thailand Research Fund (TRF) No. PHD60I0023. The authors also would like to thank Sakda MANEEROT of LAMOOL TRANSFORMERS Co., Ltd. and TESLA POWER Co., Ltd. for his technical advices.

Conflicts of Interest: The authors declare no conflict of interest.

References

1. IEC 60270. *High-Voltage Test Techniques: Partial Discharge Measurement*; International Electrotechnical Commission (IEC): Geneva, Switzerland, 2015.
2. IEEE Standard 4TM-2013. *IEEE Standard for High-Voltage Testing Techniques*; Institute of Electrical and Electronics Engineers (IEEE): Piscataway, NJ, USA, 2013.
3. IEC 60060-1. *High-Voltage Test Techniques. Part 1: General Definitions and Test Requirements*, 3rd ed.; International Electrotechnical Commission (IEC): Geneva, Switzerland, 2010.
4. IEC 61869-1. *Instrument Transformer-Part 1: General Requirements*; International Electrotechnical Commission (IEC): Geneva, Switzerland, 2007.
5. IEC 61869-3. *Instrument Transformer-Part 3: Additional Requirements for Inductive Voltage Transformers*; International Electrotechnical Commission (IEC): Geneva, Switzerland, 2011.
6. Leelachariyakul, B.; Yutthagowith, P.; Potivejkul, S. PD Detection Test of a Voltage Transformer using a Variable Frequency Converter. In Proceedings of the International Symposium on EMC and Transients in Infrastructures (ISET), Chonburi, Thailand, 26–27 November 2015.
7. Kitcharoen, P.; Kunakorn, A.; Yutthagowith, P.; Limcharoen, W. Multilevel frequency converters and noise reduction for partial discharge tests. In Proceedings of the International Conference on Electrical Engineering/Electronics, Computer, Telecommunications and Information Technology (ECTI-CON), Phuket, Thailand, 27–30 June 2017.
8. Prombud, T.; Kitcharoen, P.; Yutthagowith, P. Development of a partial discharge testing system for potential transformers. In Proceedings of the IEEE International Conference on Industrial Technology (ICIT), Lyon, France, 20–22 February 2018.
9. Yomkaew, N.; Marukatat, N.; Yutthagowith, P. A Partial Discharge Testing System Based on A 5-Level Converter with Different Control Signals. In Proceedings of the Australasian Universities Power Engineering Conference (AUPEC), Auckland, New Zealand, 27–30 November 2018.
10. Prombud, T.; Kitcharoen, P.; Yutthagowith, P. Development of a Low-Pass Filter for Partial Discharge Testing System with the Power Frequency Converter. In Proceedings of the Australasian Universities Power Engineering Conference (AUPEC), Auckland, New Zealand, 27–30 November 2018.
11. Leelachariyakul, B.; Yutthagowith, P. The Development of an Adjustable Low-pass Filter for a Partial Discharge Detection System under Testing with the Power Frequency Converter. In Proceedings of the International Universities Power Engineering Conference (UPEC), Bucharest, Romania, 3–6 September 2019.
12. Prombud, T.; Yutthagowith, P. Development of High-voltage Testing System Based on Power Frequency Converter Used in Partial Discharge Tests of Potential Transformers. *Sens. Mater.* **2020**, *32*, 573–585. [CrossRef]
13. Holmes, D.G.; Lipo, T.A. Pulse Width Modulation for Power Converters. In *Principles and Practice*; John Wiley & Sons INC.: Hoboken, NJ, USA, 2003.
14. Rashid, M.H. *Power Electronics Handbook*, 3rd ed.; Butterworth-Heinemann: Waltham, MA, USA, 2011.
15. Thiede1, A.; Martin, F. *Power Frequency Inverters for High Voltage Tests*; High-Volt Prüftechnik Dresden GmbH, High-Volt Colloquium: Dresden, Germany, 2007.
16. Hauschild, W.; Lemke, E. *High-Voltage Test and Measuring Techniques*; Springer: Cham, Switzerland, 2014.
17. User Manual for the Device MPD600, Brochure, Downloaded in January 2019. Available online: https://www.omicronenergy.com/en/products/mpd-600/documents/ (accessed on 16 January 2019).

Article

Correct Cross-Section of Cable Screen in a Medium Voltage Collector Network with Isolated Neutral of a Wind Power Plant

Huthaifa A. Al_Issa [1], Mohamed Qawaqzeh [1], Alaa Khasawneh [1], Roman Buinyi [2,*], Viacheslav Bezruchko [2] and Oleksandr Miroshnyk [3]

1   Electrical and Electronics Engineering Department, Al Balqa Applied University, 19117 Al Salt, Jordan; h.alissa@bau.edu.jo (H.A.A.); qawaqzeh@bau.edu.jo (M.Q.); alaa.khasawneh@bau.edu.jo (A.K.)
2   Department of Power Electrical Engineering, Chernihiv Polytechnic National University, str. Shevchenka, 95, 14035 Chernihiv, Ukraine; slavajm@meta.ua
3   Department of Electricity and Energy Management, Kharkiv Petro Vasylenko National Technical University of Agriculture, str. Rizdviana, 19, 61052 Kharkiv, Ukraine; omiroshnyk@khntusg.info
*   Correspondence: buinyiroman@gmail.com; Tel.: +380-462-665-126

Abstract: The article discusses the selection of cables for power lines connecting wind turbine generators at the wind power plant. The screen cross-section of these cables should be selected considering the value of the screen current at double line-to-earth fault. To calculate this current, the dimensions of the cable should be known. However, these parameters are hidden and cannot be used during designing. Therefore, a highly simplified method is currently used in practice. It is shown that the errors from the highly simplified method are up to 33%. Authors propose a simplified method based on open data of cable manufacturers. The proposed method is compared with simulation results of a common model of cable power line and takes into account self and mutual inductances of the cores and screens. It is shown that the error of the proposed method is smaller than 4.0% for real cable power lines at wind power plants. However, for a long section of cable power line (2.5 km) the error of calculation might increase up to 6.3%. This allows us to use the proposed method for designing. In addition, the authors show how the results of the highly simplified method can be corrected to improve accuracy.

Keywords: wind power plant; cable line; cable screen; double line-to-earth fault current

Citation: Al_Issa, H.A.; Qawaqzeh, M.; Khasawneh, A.; Buinyi, R.; Bezruchko, V.; Miroshnyk, O. Correct Cross-Section of Cable Screen in a Medium Voltage Collector Network with Isolated Neutral of a Wind Power Plant. *Energies* **2021**, *14*, 3026. https://doi.org/10.3390/en14113026

Academic Editor: Dan Doru Micu

Received: 20 April 2021
Accepted: 19 May 2021
Published: 24 May 2021

1. Introduction

Recently, the number of wind power plants (WPPs) has been growing in the Integrated Power System of Ukraine (IPS), the European Network of Transmission System Operators for Electricity, electric utility systems of North America and other countries. This is due to the fact that at the state level, the so-called "green" generation, which uses renewable energy resources, is stimulated. Of particular interest to investors are powerful WPP, consisting of many individual wind turbine generators (WTGs) with a nameplate rating ranging from 1.5 MW to 5.5 MW. For this power, doubly-fed induction generators (DFIGs) with wound rotors are used. In Ukraine generally each WTG has a step-up transformer, which is a pad-mounted unit, to connect to a medium-voltage collector network operating at 10 kV to 35 kV. The collector network consists of one or several feeders. All WTGs are divided into groups connected to different feeders. Typically, WTGs are connected in series (Figure 1). All feeders are connected together at a collector system station. As a rule, a station can have one or more transformers which step-up voltage to the rated value of the interconnection point. If the station is not adjacent to the interconnection point, an interconnection transmission line is used.

To increase the generation of electrical energy, WTGs are positioned as high as possible. Modern generators are located at a height of more than 80 m with a distance between towers of more than 200 m, which requires a large area, which is mostly used for agricultural purposes. To be able to use the land for its intended purpose, a collector network is built with underground power lines, as a rule, with cross-linked polyethylene (XLPE) insulated cables.

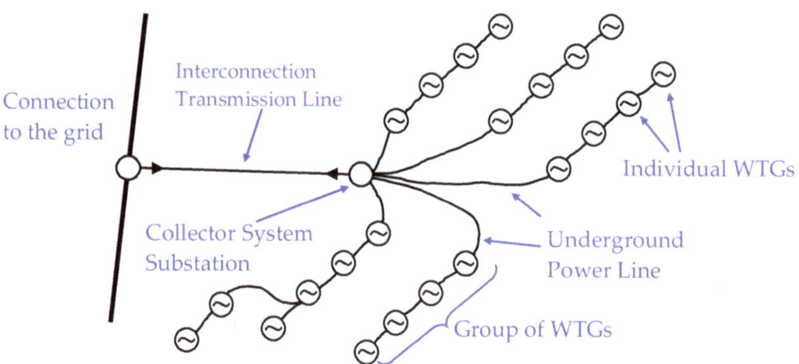

Figure 1. Wind power plant topology.

Many articles are devoted to solving a large number of problems associated with the design of cable lines with XLPE insulation, but none of them consider the case in which cable lines connect several sources of electrical energy. This is especially the case with the short length of the cable line, which we observe on the territory of the WPP.

In [1], the ways of WTG control are considered to improve the transmission capacity of a low-voltage network during a voltage dip, and in [2] a WTG control system to reduce inter-area oscillations that can be formed during parallel operation of a WPP with a power system. However, the processes considered in [1,2] will also depend on the parameters of the cable line that connects the WTG.

The selection of cross-sections of cable core and optimization of the WPP network structure by various algorithms to minimize electrical energy losses are well described in [3–8]. The articles [9,10] also raise problems of minimizing power losses in electrical networks. In [11], methods of calculating the currents that can flow through the cable screen in normal mode are considered. In [12,13], it is proposed to identify the aging of cable insulation by currents in screens that are grounded on both sides. The article [14] also raises questions of fault identification. However, these articles do not discuss the task of selecting screen cross-section of the cable. Therefore, authors raised problems of correct calculation of faults current which helps to select screen cross-section of the cable. The results give the possibility to reduce the cost of the WPPs power network.

2. Description of the Problem

The selection of cross-sections of conductors of XLPE insulated cables and their screens should be carried out in accordance with the requirements of the existing regulatory documents [15,16]. The cross-section of the cable cores is selected according to the limiting continuous current of load in normal condition and is checked by the current in the post-fault condition and also checked by the thermal short-circuit strength. In Ukraine, Italy, Finland, the Russian Federation and other countries medium voltage power networks (10–35 kV) operate with isolated (undergrounded) or resonant grounded (using grounded-fault neutralizer) neutral. Using this neutral mode allows the network to continue operating after a line-to-earth fault, which increases reliability of power generation, but in some countries, operation of the network after a line-to-earth fault is unacceptable. In the article, we only considered networks where this operation is allowed.

As known, cables with XLPE insulation are very reliable, and a line-to-earth fault in networks built using this type of cable occurs most often in the cable joint (end) sleeve. In these networks during a line-to-earth fault the line-to-screen voltage increases to a line-to-line value. This fact increases the possibility of other faults. If a second line-to-earth fault occurs on one of the other phases of the cable line, much higher current will flow from line to line through the cable screen (interphase fault current through the cable screens or double line-to-earth fault current). Therefore, when choosing the cross-section of the cable screen, it is necessary to take into account the fault like this. Because the thermal effect of this current can damage the cable.

The line-to-earth faults rarely occur in cable but more often occur in the cable joint (or end) sleeve [17]. In the worst case, the double line-to-earth fault can occur simultaneously in the cable joint (or end) sleeve on the left and right side in different phases of one cable line section (between two WTGs). In this case, the current value will be the highest. Hereinafter we will call this current the double line-to-screen fault current.

Figure 2 shows a single-line diagram in the worst case when line-to-earth faults occur simultaneously on the left and right side in the cable line section between two WTGs.

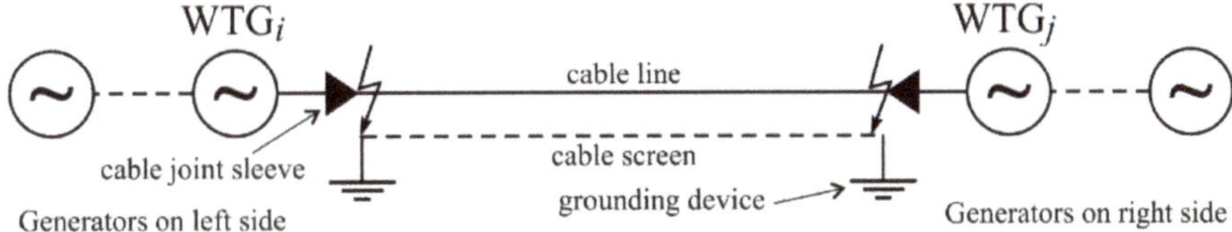

Figure 2. Schematic single-line diagram from some section of the cable line, which connects two WTGs in a group.

We will assume that the screen of a three-core cable (or three single-core cables) is grounded on both sides. It should be noted that in the case of grounding the screens of three single-core cables on only one side, the entire double line-to-screen fault current will flow through the screen of only one core, which will require a significant increase in the cross-section of the screens. Therefore, this design is not used in practice.

The problem is that to select the screen cross-section of the cable you need to know the value of the screen fault current, but to calculate the current you need to know the screen cross-section impedance, which is not yet selected. Therefore, this problem is iterative and can be solved in two or more iterations steps:

- In the first step, for the smallest possible screen cross-section (selected earlier), the value of double line-to-screen fault current is calculated and the required screen cross-section is selected again;
- In the second step, for the selected screen cross-section, the updated value of double line-to-screen fault current I_k is calculated and the selected cross-section is checked for thermal strength.

Condition to check is:

$$I_k \leq I^{(t)}_{perm.short-circuit.screen} \tag{1}$$

where $I^{(t)}_{perm.short-circuit.screen}$ is the value of the permissible short-circuit current of the screen with the duration of a clearance time.

This current is calculated under the assumption that during a fault the temperature of the cable screen must not exceed 350 °C. For this, cable manufacturers indicate the permissible one-second screen current $I^{(1s)}_{perm.short-circuit.screen}$ with the possibility of subsequent adjustment of its value for another short-circuit clearance time according to the formula [15,16]:

$$I^{(t)}_{perm.short-circuit.screen} = \frac{I^{(1s)}_{perm.short-circuit.screen}}{\sqrt{t}} \tag{2}$$

where t is the maximum permissible short-circuit clearance time in seconds.

If the calculated permissible current of the cable screen $I^{(t)}_{perm.short-circuit.screen}$ is less than the value of I_k (condition (1) is not satisfied), then it is necessary to select the larger screen(s) cross-section and repeat the second step.

It is necessary to repeat the second step until condition (1) is satisfied.

There is a problem for engineers to calculate current I_k in this algorithm. To calculate the current, a full three-phase circuit of cable power line is needed. In this circuit, parameters such as alternating current (AC) resistance of cores (screens) and self and mutual inductance between core or other core (and screens) are applied. To calculate these parameters for the model we need to know dimensions of XLPE-cable layers (Figure A3) and geometry of the cable power line (Figure A4). However, the real dimensions of the cable layers are hidden for the engineers and they are not shown in the catalogs of cable manufacturers. So, it is difficult to use this calculation in real life. Therefore, engineers use a highly simplified method to calculate screen current. This method takes into account only sources which are located on the side of the interconnection point. This method provides acceptable results only for the case of low-power WPPs, but for high-power WPPs it leads to large errors. To avoid the negative effect of those errors, project engineers deliberately overestimate the cable cross-section using the expert method. As a result, the selected cross-section of the cable screen could be in several times larger than the required cross-section. This leads to an increase in the cost of the WPPs cable networks.

We propose a simplified calculation method and compare the calculations results with the existing methods. It should be noted that the proposed method uses open data of XLPE-cable manufacturers.

3. Methodology

We replace WTGs with equivalent sources, which have its own symmetrical three-phase fault power $S''_{k\,l}$ (or symmetrical three-phase fault current $I''_{k\,l}$) for generators on the left side and $S''_{k\,r}$ (or $I''_{k\,r}$) for generators on the right side (see Figure 3). It is possible because symmetrical three-phase fault power or current is known on the moment of selection screen cross-section. These sources have their own internal impedances $Z_{k\,l}$ and $Z_{k\,r}$.

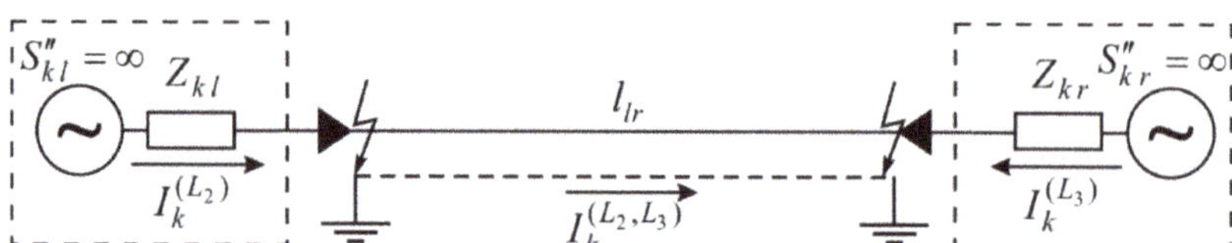

Figure 3. Single-line calculation diagram with two points at which line-to-earth faults occur simultaneously.

In the calculations, we will choose arbitrary phases in which faults occur. On the left side—in phase L_2 (current $I_k^{(L_2)}$)—and on the right side—in phase L_3 (current $I_k^{(L_3)}$). Forasmuch the impedance of the cable copper screen is much less than the total grounding impedances of the two grounding devices, then in further considerations it is assumed that all the fault current flows through the cable screen.

As a rule, only the root mean square (RMS) values of symmetrical three-phase fault currents $\left| I''_{k\,l} \right|$ and $\left| I''_{k\,r} \right|$ are known, then, according to [18], the internal impedances can be calculated by the formulas:

$$\left. \begin{aligned} Z_{k\,l} &= \frac{c \cdot U_n}{\sqrt{3} I''_{k\,l}} = r_{k\,l} + j x_{k\,l}; \\ Z_{k\,r} &= \frac{c \cdot U_n}{\sqrt{3} I''_{k\,r}} = r_{k\,r} + j x_{k\,r}, \end{aligned} \right\} \tag{3}$$

where c—voltage factor;

U_n—rated line voltage;

$r_{k\,l}, r_{k\,r}, x_{k\,l}, x_{k\,r}$—equivalent resistive and imaginary impedances of the power sources on left and right side, and modulus of impedance of such sources according to the formulas:

$$\left.\begin{aligned}
|Z_{k\,l}| &= \frac{c \cdot U_n}{\sqrt{3} \cdot |I''_{k\,l}|} = \sqrt{r^2_{k\,l} + x^2_{kl}} = \sqrt{r^2_{k\,l} + (n_l \cdot x_{k\,l})^2} = r_{k\,l} \cdot \sqrt{1 + n^2_l}; \\
|Z_{k\,r}| &= \frac{c \cdot U_n}{\sqrt{3} \cdot |I''_{k\,r}|} = \sqrt{r^2_{k\,r} + x^2_{k\,r}} = \sqrt{r^2_{k\,r} + (n_r \cdot x_{k\,r})^2} = r_{k\,r} \cdot \sqrt{1 + n^2_r},
\end{aligned}\right\} \tag{4}$$

where n_l, n_r—the ratio between the imaginary and resistive impedances of the left and right equivalent sources.

Using the ratio between the imaginary and resistive impedances, we can calculate the resistive impedances of the sources using the formulas:

$$\left.\begin{aligned}
r_{k\,l} &= \frac{|Z_{k\,l}|}{\sqrt{1+n^2_l}}; \\
r_{k\,r} &= \frac{|Z_{k\,r}|}{\sqrt{1+n^2_r}},
\end{aligned}\right\} \tag{5}$$

and imaginary impedances as $x_{k\,l} = r_{k\,l} \cdot n_l$, $x_{k\,r} = r_{k\,r} \cdot n_r$.

The above assumptions allowed us to obtain a three-phase equivalent circuit of the part of the electrical network between two WTGs (Figure 4). This circuit allows to calculate the double line-to-screen fault current I_k which flows through the cable screen. In the figure, the impedance of the cable section is given by resistive r_{core} and imaginary x_{core} impedances of a cable core and resistive impedances of a cable screen resistances r_{scr}. These parameters can be calculated from catalog data of cable manufacturers considering the length of the cable power line.

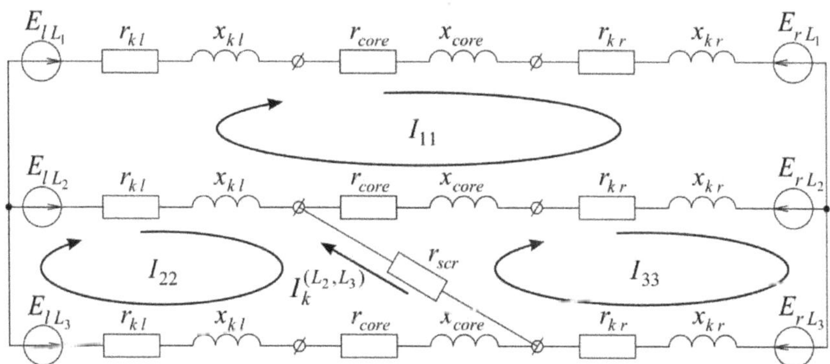

Figure 4. Equivalent circuit of the part of electrical network between two WTGs to calculate the double line-to-screen fault current with given directions of the mesh-currents.

Equivalent sources have symmetrical phase electromotive force (EMF) of the positive sequence:

$$\left.\begin{aligned}
E_{l\,L_1} &= \tfrac{c \cdot U_{nl}}{\sqrt{3}} \cdot e^{j0°}; \quad E_{r\,L_1} = \tfrac{c \cdot U_{nr}}{\sqrt{3}} \cdot e^{j\phi°}; \\
E_{l\,L_2} &= \tfrac{c \cdot U_{nl}}{\sqrt{3}} \cdot e^{-j120°}; \quad E_{r\,L_2} = \tfrac{c \cdot U_{nr}}{\sqrt{3}} \cdot e^{-j(120-\phi)°}; \\
E_{l\,L_3} &= \tfrac{c \cdot U_{nl}}{\sqrt{3}} \cdot e^{j120°}; \quad E_{r\,L_3} = \tfrac{c \cdot U_{nr}}{\sqrt{3}} \cdot e^{j(120+\phi)°},
\end{aligned}\right\} \tag{6}$$

where U_{nl}, U_{nr}—line-to-line voltages of the left and right sources which are equal to the rated voltages of the network, $U_{nl} = U_{nr} = U_n$, ϕ—the phase angle between phase EMF of the left and the right source.

The cable screen cross-section should be selected for the case of maximum current. As shown in Appendix A the current will be maximum if the phase angle is equal to zero, because in the next calculation the phase angle between phase EMF of the left and the right source is taken as equal to zero.

For single-core cables in which the screens are grounded on both sides, the screen impedance is equal to the equivalent impedance of the three screens of the cables connected in parallel, $r_{scr}/3$.

The value of the double line-to-screen fault current in the circuit, shown on Figure 4, can be calculated by the mesh-current method [19].

Loop mesh-current equation for the equivalent circuit:

$$\left.\begin{array}{l} I_{11} \cdot 2 \cdot (Z_{k\,l} + Z_{k\,r} + Z_{core}) - I_{22} \cdot Z_{k\,l} - I_{33} \cdot (Z_{k\,r} + Z_{core}) = E_{l\,L_1} - E_{l\,L_2} - E_{r\,L_1} + E_{r\,L_2}; \\ -I_{11} \cdot Z_{k\,l} + I_{22} \cdot (2 \cdot Z_{k\,l} + Z_{core} + r_{scr}) - I_{33} \cdot r_{scr} = E_{l\,L_2} - E_{l\,L_3}; \\ -I_{11} \cdot (Z_{k\,r} + Z_{core}) - I_{22} \cdot r_{scr} + I_{33} \cdot (2 \cdot Z_{k\,r} + Z_{core} + r_{scr}) = -E_{r\,L_2} + E_{r\,L_3}. \end{array}\right\} \tag{7}$$

Double line-to-screen fault current, calculated by the formula:

$$I_k^{(L_2,L_3)} = I_{33} - I_{22}. \tag{8}$$

In the case of using single-core cable in which cable screens are grounded on both sides, double line-to-screen fault current is calculated as:

$$I_{scr} = \left| I_k^{(L_2,L_3)} \right| / 3. \tag{9}$$

4. Results of Analytical Calculation for Real WPP

For example, consider the real calculation of double line-to-screen fault current for the part of electrical network of Sivash WPP (which is located in the south of Ukraine) with total power 250 MW between two WTGs, where a cable $3 \times$ NA2XS(FL)2Y $1 \times 400/35$ (TELE-FONIKA Kable S.A., Myślenice, Poland) with XLPE insulation and length equal 0.42 km is used. This cable line has resistive and imaginary impedances of the cores $Z_{core} = 0.033 + j0.046\ \Omega$ and resistive impedances of the screen $r_{scr} = 0.223/3\ \Omega$. These parameters were taken from catalogs of the cable manufacturer [20].

We assume that EMFs in phases of equivalent circuit (Figure 4) is three-phase symmetrical $35/\sqrt{3}$ kV. Equivalent resistive and imaginary impedances of the sources were calculated from the known value of symmetrical three-phase fault current $I''_{k\,l}$ and $I''_{k\,r}$ by Equations (3)–(5). To simplify the calculations, it is assumed that the sources have only imaginary impedances $Z_{k\,l} = j1.622\ \Omega$ and $Z_{k\,r} = j8.3\ \Omega$.

Solved loop mesh-current Equation (7) for the equivalent circuit are:

$$\left\{\begin{array}{l} I_{11} = 0.02012 - j0.01328\ \text{kA}; \\ I_{22} = -10.61409 - j0.40050\ \text{kA}; \\ I_{33} = 2.11053 + j0.05431\ \text{kA}. \end{array}\right. \tag{10}$$

Double line-to-screen fault current through three cable screens:

$$I_k^{(L_2,L_3)} = 12.72462 + j0.45482\ \text{kA}, \tag{11}$$

and magnitude of current $I_k^{(L_2,L_3)} = 12.73$ kA. Double line-to-screen fault current through one cable screen calculated in Equation (9) equals to 4.24 kA.

5. Results Simulation in the MATLAB Simulink for Real WPP

The obtained analytical values of the double line-to-screen fault current were verified by modeling in the MATLAB Simulink R2015b (MathWorks, Natick, MA, USA) (see Figure 5) using a common model of cable power line. The model was built considering the mutual inductances of the cores and screens of the cable power line. The impedance of the grounding devices were taken to be 10 Ω. The internal structure of block "3 Cable with screen" is shown in Appendix B.

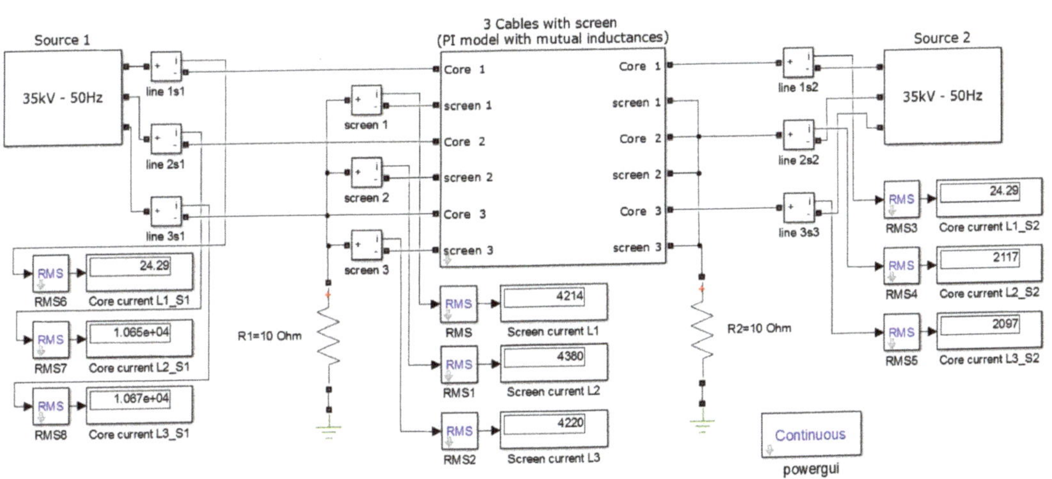

Figure 5. Model in MATLAB Simulink for calculating the current through the cables screen during a double line-to-earth fault in the case of cable lines laid with single-core cables in which the screens are grounded on both sides and taking into account two energy sources.

The parameters for the model of the cable power line with used cable type $3\times$NA2XS(FL)2Y $1\times400/35$ were obtained by formulas in Appendix C. For this type of cable dimensions (see Figure A3) are $d_{oc} = 23.2$ mm, $d_{is} = 42.2$ mm, $d_{os} = 43.8$ mm, $D_{cable} = 53.6$ mm; self-resistances of core and screen conductor are $R_c = 0.07$ Ω/km, $R_s = 0.5$ Ω/km. For the earth resistivity $\rho_e = 100$ $\Omega\cdot$m and frequency $f = 50$ Hz, self- and mutual inductances were obtained: $L_c = 2.309$ mH/km, $L_s = 2.134$ mH/km, $L_{c-s} = 2.135$ mH/km, $L_m = 1.953$ mH/km. These dimensions of the cable layers are not included in the catalogs of cable manufacturers, that is why they are not known to engineers. As a result, it is difficult to use this calculation in real life.

For the XLPE core-screen insulated ($\varepsilon_{c-s} = 2.3$) and PE screen-earth insulated ($\varepsilon_{s-e} = 2.3$) the capacitance between the core and screen $C_{c-s} = 0.214 \cdot 10^{-6}$ F/km and capacitance between the screen and earth $C_{s-e} = 0.634 \cdot 10^{-6}$ F/km.

Figure 5 shows that the currents in the three screens of cables during the double line-to-screen fault are 4.214 kA, 4.38 kA, 4.22 kA, which almost coincides with the value obtained analytically—4.24 kA. This confirms the adequacy of the proposed principle of analytical calculation of the current through the screen(s) of cables during a double line-to-earth fault. However, it is necessary to analyze in detail the error of calculation for other cable power lines parameters.

6. Results of Analysis of Calculation Accuracy of the Current

Here we analyze the accuracy of our simplified analytical method. We calculate the relative error from the simulation based on the common model of the power line as:

$$\delta_{Li} = \frac{I'^{(Li)}_{scr} - I_{scr}}{I'^{(Li)}_{scr}} \cdot 100\% \tag{12}$$

where $I'^{(Li)}_{scr}$—cable screen currents in i-phase (L_1, L_2 and L_3) obtained using the simulation in MATLAB Simulink; I_{scr}—cable screen currents obtained using the proposed simplified analytical method.

Figure 6a shows the dependence of the error δ on cable screen cross-section from 16 to 35 mm². The dependence was plotted for the cable mentioned above with length 1 km and core cross-section 400 mm² and was added a range of error variation when changing length of cable line from 0.5 to 2.5 km. In the figure it is shown relative error of current calculation in screen L_1 (red), L_2 (blue) and L_3 (green). From the figure it is seen that for the proposed simplified analytical method the error does not exceed 6.3% (this maximum value of error is for the case of the maximum cross-section of the cable screen which is applied in WPPs and maximum length). Generally, on the real WPPs the WTGs are located with distance from 0.5 to 1 km and the calculated cable screen cross-section of one line is 25 mm². In this case of real WPPs the maximum error will be only 3.3%. However, the error can be different other core cable cross-sections.

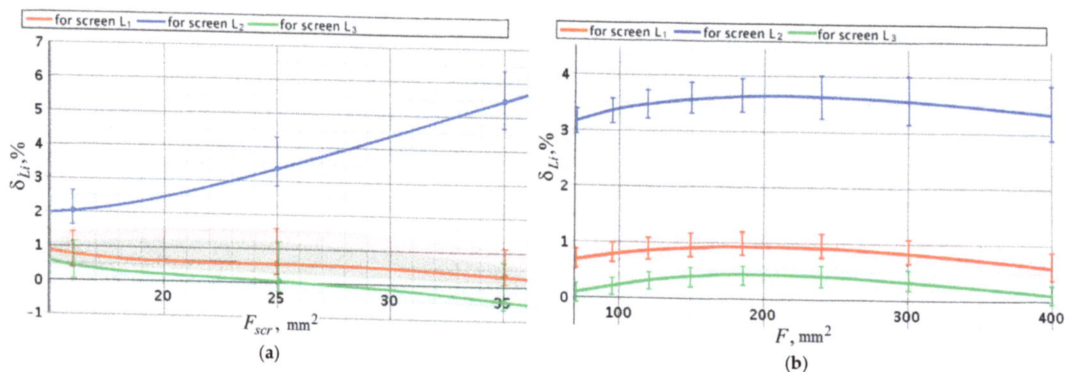

Figure 6. Dependences of the error δ on cross-section of cable screen (**a**) and cross-section of cable core (**b**) for three lines: L_1 (red), L_2 (blue) and L_3 (green).

Figure 6b shows the dependence of the error δ on cable core cross-section from 70 to 400 mm². The dependence was plotted for above cable with length 1 km and cable screen cross-section 25 mm² and was added a range of error variation when changing cable screen cross-section from 16 to 35 mm². From the figure it is seen that the error almost does not depend on the cable core cross-section.

As we can see for real WPPs the error will be smaller than 4.0 %. This allows to apply the proposed principle of analytical calculation to calculate the maximum current in the screen of cable during a double line-to-earth fault.

Generally, it is difficult for engineers to consider all factors that influence the current. Additionally, the calculation becomes complicated if it takes into account sources on both sides. As a rule, real calculations are made ignoring the smaller power source (highly simplified method). Therefore, we conducted research on the error of this method.

The ratio of the symmetrical three-phase fault currents on the buses of each of the sources (WTGs) $I''_{k\,l}$ and $I''_{k\,r}$ has the greatest influence on the value of the calculated current. Analysis of the values of symmetrical three-phase fault currents on the buses of 10–35 kV electrical power networks in Ukraine ($I''_{k\,l}$) and similar fault currents on the buses of wind turbine groups ($I''_{k\,r}$) showed that the ratio of currents $k = I''_{k\,l}/I''_{k\,r}$ is within the range from 3 to 10. Therefore, we plotted the dependence of the ratio between the value of the calculated current obtained using the simulation in MATLAB Simulink and highly simplified method (with two and one source $n = I^{(2source)}_{scr}/I^{(1source)}_{scr}$) only for this range of k (Figure 7). Figure 7 also shows the domain in which n will change with the change in other parameters: the length of power line (from 0.5 to 2.5 km), cable screen cross-section (from 16 to 35 mm^2) and cable core cross-section (from 70 to 400 mm^2).

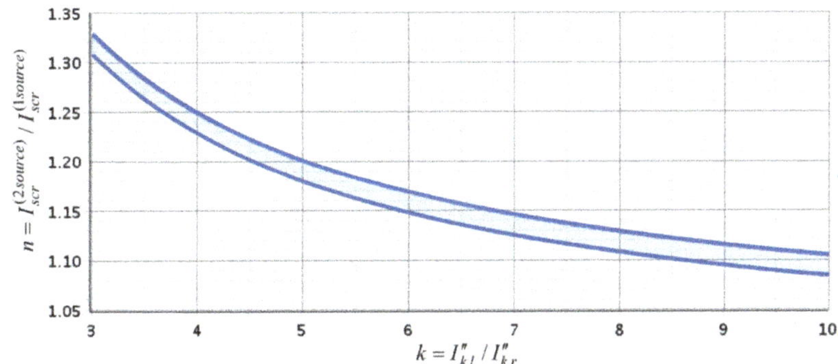

Figure 7. Dependence of the ratio between the value of the current obtained using the simulation in MATLAB Simulink and the value calculated using the highly simplified method on the ratio of the symmetrical three-phase fault currents on the buses of each sources (WTGs).

From the chart shown in Figure 7, it can be seen that the error of calculation will be in the range from 33 to 10 %. The domain of variability of n is narrow.

This indicates the need to take into account the second source when calculating the double line-to-screen fault current in order to further select the cable screen.

Based on the above, we can recommend to use the highly simplified method which takes in account only one source (the most powerful). However, the result of calculation should be multiplied on the ratio n from the chart in Figure 7. In this case engineers have to calculate the ratio of the symmetrical three-phase fault currents on the buses of each sources $k = I''_{k\,l}/I''_{k\,r}$, and use this value to find n from the chart in Figure 7.

7. Conclusions

The objective of this paper was to improve the accuracy of calculation the cross-section of cable screen in a medium voltage collector network with isolated neutral used in wind power plants. Currently in real setting a highly simplified method is used which leads to large errors. To avoid the negative effects of such errors, project engineers have to choose a larger cross-section of the cable screen using an expert method. Thus, the selected cross-section of the cable screen is often several times larger than the required cross-section. This leads to an increase in the cost of the WPPs cable networks.

The results obtained in this paper can be summarized as follows:

1. The simplified analytical method was proposed for calculating cable screen current during double line-to-earth fault on a cable line section of a medium voltage network with isolated neutral of a wind power plant. The advantage of the proposed method is that the method uses open data from catalogs of cable manufacturers. In contrast, the common model requires the dimensions of XLPE-cable layers which are hidden;

2. Results of calculation of the proposed simplified method were compared with the simulation results of a common model of cable power line that takes into account self- and mutual inductances of the cores and screens in different phases. It is shown that for parameters of the real WPPs (the length of section power line 0.5–2.5 km, cable screen cross-section 16–25 mm^2 and cable core cross-section 70–400 mm^2) the maximum error of the proposed method will be smaller than 4.0%. This allows to use the proposed method for design;

3. The analysis of calculation results of a highly simplified method (which is currently used in practice) shows that its relative error can be in the range from 10 to 33% from the common model simulation results. It was shown that the error of calculation is

little dependent on the cable line parameter of the real WPPs (the length of section, screen and core cross-section of cable line) but the error is highly dependent on the ratio of the symmetrical three-phase fault currents on the left and right side of cable power line section. Therefore, to simplify the calculation in practice we obtained dependence which allows to correct the results of the highly simplified method. We recommend to use this method only with the correction according to Figure 7.

Author Contributions: Conceptualization, R.B. and M.Q.; methodology, H.A.A. and R.B.; validation, M.Q., H.A.A., A.K., O.M. and V.B.; formal analysis, R.B.; investigation, R.B.; resources, R.B.; writing—original draft preparation, R.B.; writing—review and editing, R.B. and V.B. All authors have read and agreed to the published version of the manuscript.

Funding: This research received no external funding.

Institutional Review Board Statement: Not applicable.

Informed Consent Statement: Not applicable.

Data Availability Statement: The data presented in this study are available on request from the corresponding author.

Conflicts of Interest: The authors declare no conflict of interest.

Appendix A

To control the power supplied by the generator, the method of changing the angle of the generator EMF relative to the EMF of the external grid can be used. We analyzed the effect of the angle on the screen current during double line-to-screen fault.

Dependencies of the double line-to-screen fault current I_{scr} from the phase angle ϕ between phase EMF of left and right sources are shown in Figure A1.

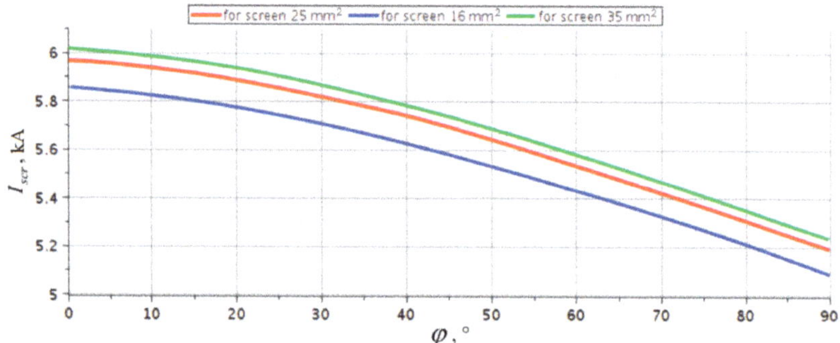

Figure A1. Dependence of the double line-to-earth fault current through the cable screen(s) I_{scr} from the phase angle φ between phase EMF of the left and the right sources ($U_{nl} = U_{nr} = 35$ kV; $I''_{k\,l} = 20$ kA; $n_l = 10$; $I''_{k\,r} = 2.4$ kA; $n_r = 3$; $F = 400$ mm²; $l = 1$ km).

Figure A1 shows that with an increase in the angle from 0 to 90 degrees, the current decreases. For other input parameters the shape of the curve will stay the same with the maximum at zero angle. This indicates that the calculation condition should be held for the angle equal to zero because the selection of cable screen cross-section should be done for case of maximum current.

Appendix B

In the MATLAB Simulink we created a model of cable power line with Mutual Inductance block (Figure A2). This model takes into account the mutual inductance between cores and screens with earth return, capacitance between core and screen and the screen and earth.

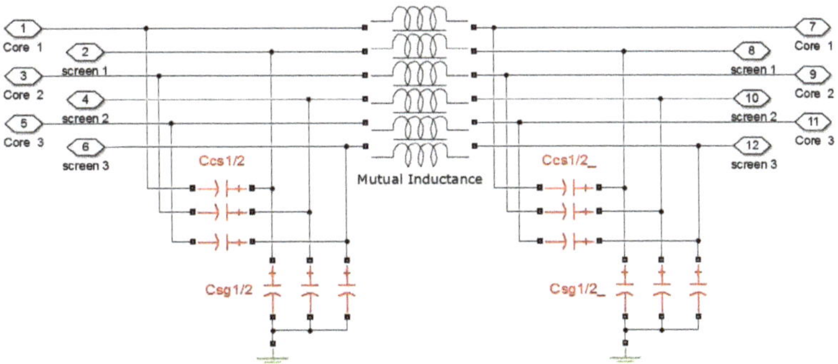

Figure A2. Model of cable power line.

Parameters of model is given by:

$$R = \begin{bmatrix} R_{c1} & 0 & 0 & 0 & 0 & 0 \\ 0 & R_{s1} & 0 & 0 & 0 & 0 \\ 0 & 0 & R_{c2} & 0 & 0 & 0 \\ 0 & 0 & 0 & R_{s2} & 0 & 0 \\ 0 & 0 & 0 & 0 & R_{c3} & 0 \\ 0 & 0 & 0 & 0 & 0 & R_{s3} \end{bmatrix} ; \; L = \begin{bmatrix} L_{c1} & L_{c1-s1} & L_m & L_m & L_m & L_m \\ L_{c1-s1} & L_{s1} & L_m & L_m & L_m & L_m \\ L_m & L_m & L_{c2} & L_{c2-s2} & L_m & L_m \\ L_m & L_m & L_{c2-s2} & L_{s2} & L_m & L_m \\ L_m & L_m & L_m & L_m & L_{c3} & L_{c3-s3} \\ L_m & L_m & L_m & L_m & L_{c3-s3} & L_{s3} \end{bmatrix} ; \qquad \text{(A1)}$$

$$C = \begin{bmatrix} C_{c1-s1} & -C_{c1-s1} & 0 & 0 & 0 & 0 \\ -C_{c1-s1} & C_{s1-e} & 0 & 0 & 0 & 0 \\ 0 & 0 & C_{c2-s2} & -C_{c2-s2} & 0 & 0 \\ 0 & 0 & -C_{c2-s2} & C_{s2-e} & 0 & 0 \\ 0 & 0 & 0 & 0 & C_{c3-s3} & -C_{c3-s3} \\ 0 & 0 & 0 & 0 & -C_{c3-s3} & C_{s3-e} \end{bmatrix} . \qquad \text{(A2)}$$

The parameters of the model of cable line are obtained by formulas in Appendix C.

Appendix C

The appendix contains a description of the calculation of electrical cable power line parameters. To calculate these parameters, we have to know the geometric dimensions of the cable. Simplified image of the XLPE-cable structure is shown in Figure A3.

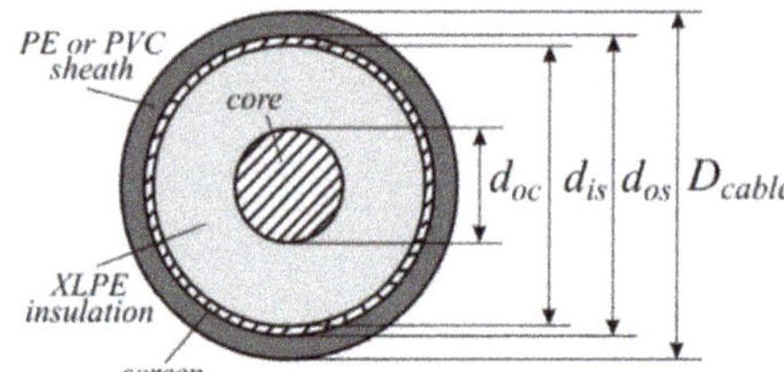

Figure A3. Simplified image of the XLPE-cable layers structure.

However, real dimensions of the XLPE-cable layers are hidden for the engineers and are not included in the catalogs of cable manufacturers. So, it is difficult to use this calculation in real life.

The self-inductance of a cable core conductor can be obtained from the formula [21]:

$$L_c = \frac{\mu_0}{2\pi} \cdot \ln\left(\frac{2D_{erc}}{d_{oc}} + \frac{1}{4}\right),$$
(A3)

where μ_0 is the permeability of free space (or air) equals to $4\pi \cdot 10^{-4}$ H/km; D_{erc} is the depth of equivalent earth return conductor is given by $D_{erc} = 658,87\sqrt{\frac{\rho_e}{f}}$ in meters; ρ_e is the earth resistivity in Ω·m; f is the current frequency in Hz.

The self-inductance of a copper screen:

$$L_s = \frac{\mu_0}{2\pi} \cdot \ln\left(\frac{2D_{erc}}{d_{os}} + \frac{1}{4} \cdot \left(1 - \frac{2d_{is}^2}{d_{os}^2 - d_{is}^2} + \frac{4d_{is}^4}{\left(d_{os}^2 - d_{is}^2\right)^2} \cdot \ln\left(\frac{d_{os}}{d_{is}}\right)\right)\right).$$
(A4)

The mutual inductance between core or screen with earth return, is given by:

$$L_{c-s} = \frac{\mu_0}{2\pi} \cdot \ln\left(\frac{D_{erc}}{GMD}\right),$$
(A5)

where *GMD* is the geometric mean distance between the two conductors, e.g., the *GMD* between the core and the screen of cable is given by:

$$GMD = \frac{1}{2} \cdot \left(d_{is} + \frac{d_{os} - d_{is}}{2}\right).$$
(A6)

To calculate mutual inductances, we should choose the type of three single-core cable layouts. In our case the trefoil formation layout is chosen (see Figure A4).

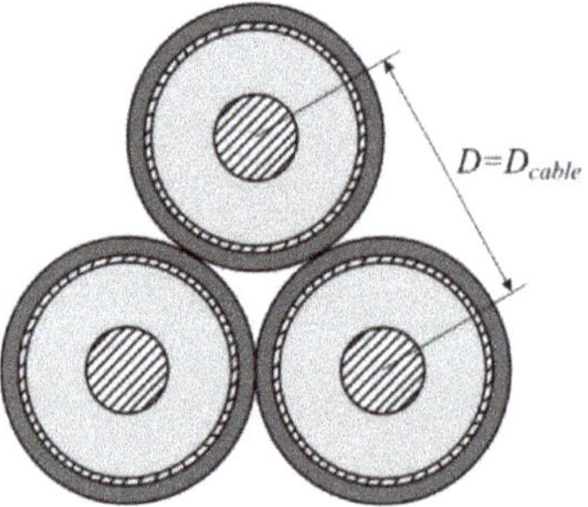

Figure A4. Single-core cable in the trefoil formation.

The mutual inductance between core and other core (core and screen) with earth return, is given by:

$$L_m = \frac{\mu_0}{2\pi} \cdot \ln\left(\frac{D_{erc}}{D}\right),\tag{A7}$$

where D is the *GMD* between the centers of cables. For a trefoil formation cable power line $D = D_{cable}$.

The capacitance between the core and screen is given by:

$$C_{c-s} = \frac{0.0556325 \cdot \varepsilon_{c-s}}{\ln\left(\frac{d_{is}}{d_{oc}}\right)} \cdot 10^{-6}, \text{ F/km}\tag{A8}$$

Similarly, the capacitance between the screen and earth is given by:

$$C_{s-e} = \frac{0.0556325 \cdot \varepsilon_{s-e}}{\ln\left(\frac{D_{cable}}{d_{os}}\right)} \cdot 10^{-6}, \text{ F/km}\tag{A9}$$

where ε_{c-s} and ε_{s-e} are the relative permittivities of core-to-screen and screen-to-earth insulation.

References

1. Nadour, M.; Essadki, A.; Nasser, T. Improving low-voltage ride-through capability of a multimegawatt DFIG based wind turbine under grid faults. *Prot. Control. Mod. Power Syst.* **2020**, *5*, 33. [CrossRef]
2. Tummala, A.S.L.V. A robust composite wide area control of a DFIG wind energy system for damping inter-area oscillations. *Prot. Control. Mod. Power Syst.* **2020**, *5*, 25. [CrossRef]
3. Semków, M. Optimisation of Cable Cross in Medium Voltage Networks of a Wind Farm. *Acta Energetica* **2013**, *17*, 132–138. [CrossRef]
4. Wang, L.; Wu, J.; Tang, Z.; Wang, T. An Integration Optimization Method for Power Collection Systems of Offshore Wind Farms. *Energies* **2019**, *12*, 3965. [CrossRef]
5. González, J.S.; Payán, M.B.; Santos, J.M.R.; González-Longatt, F. A review and recent developments in the optimal wind-turbine micro-siting problem. *Renew. Sustain. Energy Rev.* **2014**, *30*, 133–144. [CrossRef]
6. Quinonez-Varela, G.; Ault, G.; Anaya-Lara, O.; McDonald, J. Electrical collector system options for large offshore wind farms. *IET Renew. Power Gener.* **2007**, *1*, 107. [CrossRef]
7. Pillai, A.; Chick, J.; Johanning, L.; Khorasanchi, M.; De Laleu, V. Offshore wind farm electrical cable layout optimization. *Eng. Optim.* **2014**, *47*, 1689–1708. [CrossRef]
8. González, J.S.; Gonzalez-Rodriguez, A.G.; Mora, J.C.; Payán, M.B.; Santos, J.R. Overall design optimization of wind farms. *Renew. Energy* **2011**, *36*, 1973–1982. [CrossRef]
9. Buinyi, R.; Krasnozhon, A.; Zorin, V.; Kvytsynskyi, A. Justification for use of voltage class 20 kv in urban electrical networks. *Tekhnichna Elektrodynamika* **2019**, *2019*, 68–71. [CrossRef]
10. Krasnozhon, A.V.; Buinyi, R.O.; Pentegov, I.V. Calculation of active power losses in the grounding wire of overhead power lines. *Tekhnichna Elektrodynamika* **2016**, *2016*, 23–25. [CrossRef]
11. Grinchenko, V.S.; Tkachenko, O.O.; Grinchenko, N.V. Improving calculation accuracy of currents in cable shields at double-sided grounding of three-phase cable line. *Electr. Eng. Electromechanics* **2017**, *2*, 39–42. [CrossRef]
12. Li, L.; Yang, Z.; Luo, Z.; Liu, K. Transient Disturbances Based Non-Intrusive Ageing Condition Assessment for Cross-Bonded Cables. *IEEE Access* **2020**, *8*, 176651–176660. [CrossRef]
13. *IEEE Guide for Bonding Shields and Sheaths of Single-Conductor Power Cables Rated 5 kV through 500 kV*; IEEE Std 575-2014 (Revision of IEEE Std 575-1988); IEEE: New York, NY, USA, 2014; pp. 1–83. [CrossRef]
14. Bezruchko, V.M.; Buinyi, R.O.; Strogii, A.Y.; Tkach, V.I. Application of GSM technology for identification of phase-to-ground faults in electric networks with isolated neutral and pin-type isolation. *Tekhnichna Elektrodynamika* **2018**, *2018*, 96–99. [CrossRef]
15. *Electrical Installation Regulations*; Fort: Kharkiv, Ukraine, 2017; 760p.
16. SOU-N MEV 40.1-37471933-49:2011. *Design of Cable Power Lines with Voltage up to 330 kV. Requirements (with Changes 2017)*; Ministry of Energy of Ukraine: Kyiv, Ukraine, 2017; 168p.
17. Vogelsang, R.; Sekula, O.; Nyffenegger, H.; Weissenberg, W. Long-Term Experiences with XLPE Cable Systems up to 550 kV. *9th CIGRE CIRED Conference 2009 (SC B1, Kranjska Gora, Slovenia)*. Available online: https://silo.tips/download/long-term-experiences-with-xlpe-cable-systems-up-to-550-kv#modals (accessed on 10 March 2021).
18. IEC 60909-0. *Short-Circuit Currents in Three-Phase a.c. Systems-Part 0: Calculation of Currents*; International Electrotechnical Commission (IEC): Geneva, Switzerland, 2001; p. 137.
19. Charles, K.; Matthew, N. *Fundamentals of Electric Circuits*, 5th ed.; Mc Graw Hill: New York, NY, USA, 2013; 995p.

20. Official Site of NEXANS Group. Available online: https://www.nexans.ua/ (accessed on 10 March 2021).
21. Tleis, N. *Power Systems Modelling and Fault Analysis: Theory and Practice*, 1st ed.; Newnes: Oxford, UK, 2008; 650p.

Article

Impact of Cable Configuration on the Voltage Induced in Cable Screen during Work with One-Sidedly Ungrounded Cable Screen

Aleksandra Schött-Szymczak and Krzysztof Walczak *

Department of Environmental Engineering and Energy, Faculty of Electrical Engineering,
Poznan University of Technology, 60-965 Poznan, Poland; aleksandra.schott-szymczak@put.poznan.pl
* Correspondence: krzysztof.walczak@put.poznan.pl; Tel.: +48-61-665-2581

Abstract: In the latest research, it has been proven that from the point of view of losses in a cable distribution line, the most advantageous operation is to work with two or one phase of metallic cable screen ungrounded. However, such an operation may cause changes in the network characteristics and thus the occurrence of undesirable phenomena. One of those characteristics is the overvoltages in those cable screens, which can lead to cable line damage. The simulation tests presented in this article are closely related to the unusual method of operation of the MV cable screens and their performance, and they address the question of whether in a given system ground fault overvoltages may be a significant threat to the operation of the cable. The research methods used to verify these risks are related to the simulation of the cable line operating states using the DIgSILENT PowerFactory program (DIgSILENT GmbH, Gomaringen, Germany). Overvoltage simulations were performed, taking into account changes in the network configuration, such as the method of cable screens grounding, the length of cable lines, the cross-section of the conductor and cable screen, or the method of operation of the neutral point. The results for the cable line modeled as a part of the MV network with the variables considered during the tests indicate the possible impact of the one-sidedly cable screen ungrounding on overvoltages in this cable screen. The obtained results at the level of a few kV in one-sidedly ungrounded cable screens show that the change of the configuration of the operation of these cable screens may affect the safety of the network operation.

Keywords: MV cable line; metallic cable screen; overvoltages; computer simulation

Citation: Schött-Szymczak, A.; Walczak, K. Impact of Cable Configuration on the Voltage Induced in Cable Screen during Work with One-Sidedly Ungrounded Cable Screen. *Energies* **2021**, *14*, 4263. https://doi.org/10.3390/en14144263

Academic Editor: Dan Doru Micu

Received: 15 June 2021
Accepted: 12 July 2021
Published: 14 July 2021

1. Introduction

Many years of adaptation and the multitude of possible uses of cable in the power industry result in a huge selection of cable constructions and the elements they comprise. Among the conductive elements, conductors, shields, sheaths, and metallic screens can be distinguished. Except for conductors, other components, in most cases, need to be grounded in sleeves or at cable heads [1]. According to the latest research [2], the cable screen may, under certain conditions, also be an exception. Ungrounding cable screens were originally associated with a positive phenomenon [3], which is a decrease in transmission losses. However, in the course of research related to methods of reducing these losses, the negative effects of such a solution were discovered [4]. Crucial to the new solution are problems such as changing the longitudinal parameters of the cable line or the appearance of dangerous overvoltages at the end of cable runs.

As it was mentioned in the previous paragraph, the positive effect of ungrounding the cable screen of an MV cable is to reduce transmission losses [5]. These losses result from the current flow through the conductive elements of the cable line, including the cable screen [6]. Attempts to reduce these losses also included testing the configuration of the network itself in which the cable line operates [7]. In order to reduce them, a change in the grounding configuration of the cable screen has been proposed, which will terminate in even a twofold reduction of losses [8] while maintaining the basic role of this element of the cable line. This advantage is also appreciated by electricity distributors, for which changing the configuration of cable screen operation is a non-invasive and relatively cheap method to reduce transmission losses. At the time of writing this manuscript, one of the distribution companies started to implement such a method in its MV cable networks [9].

A special case, in which the influence of changing the method of a grounding cable's metallic screen should be considered, is quasi-steady states, such as ground faults. During the ground fault, a current is induced in the metallic cable screen, the flow of which increases its potential. In the case of a cable line grounded on both its ends, this current is closed by the ground, and the ground fault overvoltages that may occur are suppressed by means of the grounding system. The situation changes when one side of the cable screen is ungrounded. Proposed network configurations assume the ungrounding of one or two phases of the cable screens in a given line sequence in order to limit the flow of currents through them. In such a system, according to the research conducted earlier [10], a significant increase in ground fault overvoltages can be noticed, which appear in ungrounded conductors due to the occurrence of a returning ground fault current. The occurrence of such phenomenon may constitute a significant disadvantage of the proposed solution; however, in the case of overvoltages, their level is also important, which may define the detailed principles of configuration changes [11]. Therefore, in order to be able to determine the appropriate network operation parameters, the level of these overvoltages should be examined in relation to the conditions that may prevail in the MV network.

The basic classification of the origin of the overvoltages phenomenon in the power system networks is their internal or external imposition [12]. Since MV cable lines are completely protected against atmospheric disturbances that may cause overvoltages of external origin, only surges resulting from internal reasons should be considered. Among these, one should distinguish overvoltages that may occur during faults, sudden changes in power demand in a given network, or during switching operations [13].

The subject of this publication is the induced voltages in a metallic cable screen as a result of a ground fault in the cable line, taking into account changes in the essential parameters of the network in which this cable works. This topic is the result of previous work on the phenomenon presented in the article [12]. This publication presents the result of field research and the corresponding simulation results on the network model made in the DIgSILENT PowerFactory program. These tests included a special case of an MV network with specific component parameters operating in the system. The results obtained for this case show that with the change of the grounding system of the cable screen, the voltage induced in these conductors significantly changes, which led to the need of extending the research on this phenomenon. Due to the nature of overvoltages and possible problems resulting from their occurrence in cable lines [13], it was decided to perform a simulation test based on the basics used and tested in the above-mentioned publication, taking into account changes in the parameters of the network and its components. Section 2 of this manuscript presents the methods and network model used to conduct the research and describes the variants of variables considered during the simulation.

2. Materials and Methods

Initiating a ground fault, which will result in overvoltages, for research purposes in a real network is hazardous and requires extensive cooperation between the group of researchers and the network operator. Therefore, studies of phenomena related to the occurrence of ground faults in MV networks are mainly performed with the use of computer simulations, with the possible use of laboratory resources. To determine the magnitude of overvoltages that may occur in MV cable lines in the event of specific operating conditions, a cable model created with the use of the Power Factory program will be used. In order to be certain that the model used for the tests will meet the conditions of the normal operation of the network, for simulation tests, a modified model used for comparative tests presented in the publication [11] was applied, The purpose of the simulations carried out is primarily to get acquainted with the voltages induced in a cable screen; therefore, the model is limited to the cable-overhead MV line connected to the network through the 110/MV power station, where the ground fault necessary for the analysis of the phenomena will be performed in the overhead part. All the constant parameters concerning the model elements are presented in Table 1, while the variables are described in the following paragraphs. All values of device parameters used to model the system are based on the catalog cards. Particularly important for the creation of the grid model are power transformers [14], MV cable lines with XLPE insulation [15], grounding transformers [16], and overhead line [17] data, which have been taken from the datasheets available online. Figure 1 also presents a substitute diagram of the discussed model, which shows the main components of the network under consideration.

Table 1. Parameters of model's elements.

Symbol	Name of Element	Parameters
T1	transformer 110/15	S_n = 16 MVA υ = 110/15 u_z = 8% Δp_{CU} = 15 MW
T2	grounding transformer 15/0.4	S_n = 1.09 MVA υ = 15/0.4 u_z = 5.5% Δp_{CU} = 1.9 kW
L1	cable line	XRUHAKXS 12/20 kV laid at a depth of 0.7 m
CS/L1	metallic cable screen	copper wire
L2	overhead line	PAS 70, 2 km
LOAD	load 1	P_1 = 2 MW Q_1 = 0.5 Mvar
R1	grounding resistance	R_E = 0.5 Ω
R2	grounding resistance	R_E = 6 Ω

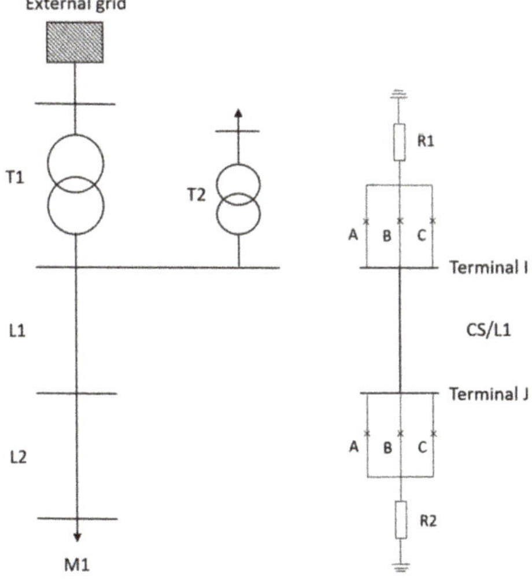

Figure 1. Diagram of network model.

According to the literature, sudden changes in the power flow through the network have a significant impact on the occurrence of overvoltages in power devices [18]. In the case considered in this manuscript, sudden changes will accompany a ground fault in the overhead line that is connected to the cable line. In connection with the adopted layout, device parameters can be distinguished in which changes will not have a major impact on the ground fault current—i.e., those that will remain constant in these considerations and those that may have a significant impact on the network parameters. The second group of these parameters will form the basis for comparative simulation studies that will allow analyzing the impact of changing these parameters on the size magnitude of overvoltages in the network. This group will include values forced by changes in the work of a neutral point grounding system, changes in the three-phase cable arrangement, changes in the cross-section of the conductor and cable screen, and changes in the cable line's length.

Knowing the model (Figure 1) and that it has been verified with previous tests, the appropriate way of conducting the research should be chosen, taking into account the tool and simulation methods. The working environment used for simulation studies was the DIgSILENT PowerFactory software for simulating the operating states of the power system and network, which includes the Simulation EMT tool, among others [19]. It enables simulation studies using the instantaneous values for electromagnetic transient states. For the proper conduct of the ground fault parameters tests, a disturbed parameter cable model was used, and the simulations were performed with an option that allows creating an independent event scenario. For all simulations performed, the event scenario included the following:

- Normal operation condition of the system lasting 0.5 s,
- Disturbance operation condition, a single phase, no-resistance ground fault in phase A of the overhead line connected to the considered cable line, lasting 0.5 s,
- Disconnection of the overhead line and disconnection of the load from the system, lasting 0.5 s.

The object of the research is the cable screens of the three-phase MV cable line and the tested parameter—the voltage appearing at the end of the cable run in individual conductors. The most common work configurations of cable screens, taking into account their grounding, are as follows:

- Single-point bonding,
- Both-ends bonding without transposition of cable screens and cores,
- Transposition of cable screen and/or cores (possible both-ends bonding or single-end bonding).

In the described research, a modification of the both-ends bonding without transposition of cable screens and cores configuration was used, which is the most advantageous from the point of view of the auxiliary role of cable screen but the most heavily burdened from the point of view of losses [20]. In order to maintain the advantages of this method of operation, it has been proposed to unilaterally unground one or two phases of the cable screen for systems with one cable line per phase. The proposed systems analyzed during the computer simulation are presented in Figure 2.

The observation of the proposed system requires the selection of appropriate variants of variables that will allow the analysis of the values in this system. The first stage of computer simulations was to compare the proposed work configurations for one-sidedly grounding of one or two phases of cable screens. At this point, it was decided to stay with the variant with only one cable screen phase ungrounded, and the remaining tests were carried out taking into account the following:

- Neutral point treatment,
- Cable arrangement,
- Cross-section of the conductor,
- Cross-section of the cable screen,
- Length of the tested cable line.

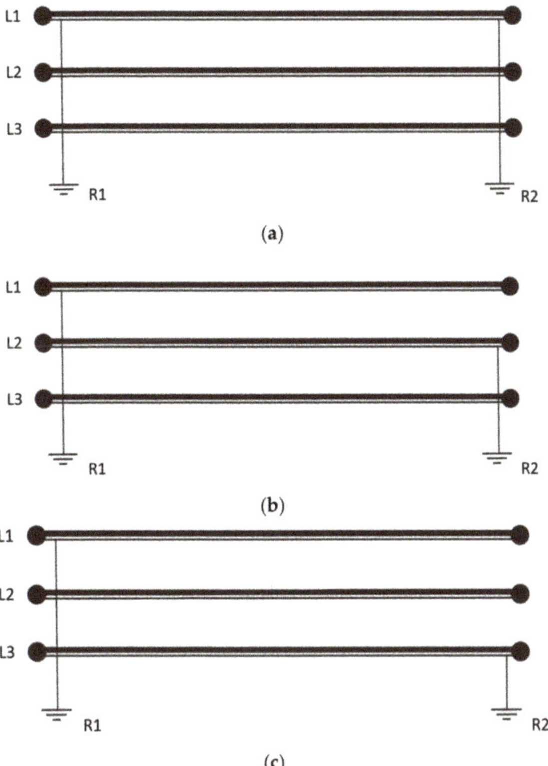

Figure 2. Cable screens grounding methods: (**a**) Both-ends bonding; (**b**) Both-ends bonding with one of the phases one-sidedly ungrounded; (**c**) Both-ends bonding with two of the phases one-sidedly ungrounded.

3. Computer Simulation Results

The following section presents the results of the simulation studies carried out with the division into previously described variables and variants. The results are presented in the form of waveforms, graphs of the maximum values of overvoltages, or comparison of maximum values, with an appropriate comment. The basic tested cable, unless otherwise stated, during the simulation is the XRUHAKXS 1 × 120/50 cable in a three-phase, triangular configuration with the length of 1 km. The summary of the outcomes and conclusions are provided at the end of the manuscript.

3.1. Cable Line Configuration

The selection of the configuration of the cable screen's work, which will ensure safe and reliable operation of the system, is possible thanks to the analysis of the overvoltages at the extreme points of the cable line. These are the places where the cable line is connected to the rest of the network in which it works and where potential transfer is possible. In the layout shown in Figure 1, it can be seen that the cable is connected to the station, which will be called Terminal I, and it is connected to the overhead line, through the cable heads, in Terminal J. Induced phase voltage in the three phases of the cable screens appearing during a single-phase ground fault I phase A of the L2 overhead line will be presented in Figures 3 and 4, respectively, for Terminal I from the station side and Terminal J from the overhead line side as voltage waveforms. The cable screens grounding configurations are also presented and compared on the voltage waveforms. The cable screen grounding methods presented in Figure 2 are applied as follows:

- Method "a"—both-ends bonding with one of the phases one-sidedly ungrounded,
- Method "b"—both-ends bonding with two of the phases one-sidedly ungrounded.

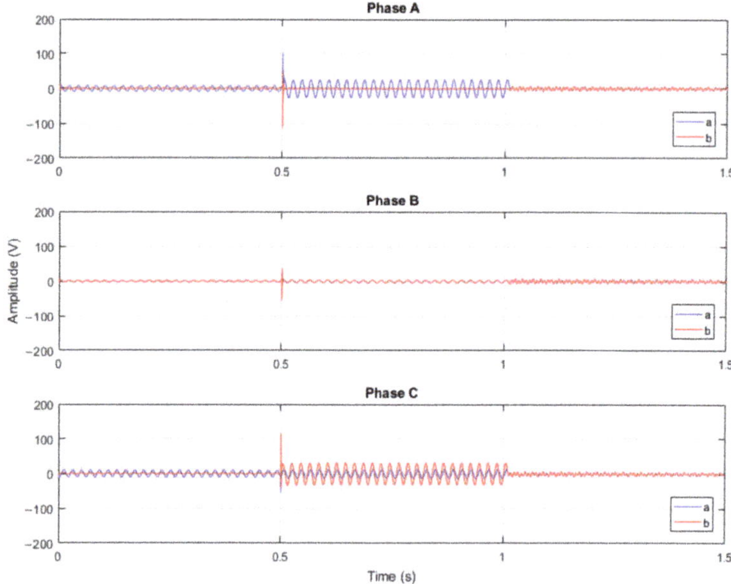

Figure 3. Voltage waveforms in the three-phase cable line screens in Terminal I.

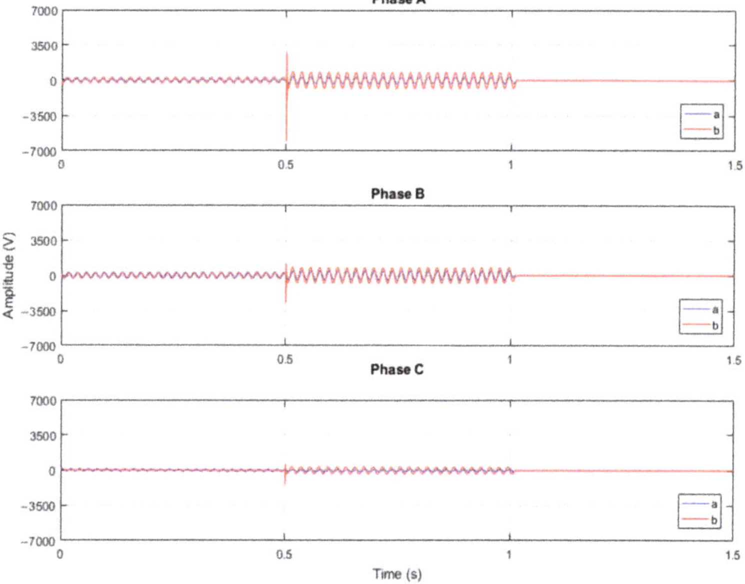

Figure 4. Voltage waveforms in three-phase cable line screens in Terminal J.

In the "a" method, the cable screen of phase A is ungrounded, while in the "b" method, the cable screen of phases A and B is ungrounded.

In Figures 3 and 4, separate voltage waveforms in three-phase cable line screens for two methods "a" and "b" are presented. The voltage values that appear during the simulation in Terminal I are significantly lower compared to Terminal J. In this situation, it is worth mentioning that the ungrounding of the cable screens takes place in Terminal J, while from the side of Terminal I, all three phases remain grounded through the R1 resistance. The ground fault current flowing from the fault location, i.e., the L2 overhead line, also flows through the cable line to the station from the Terminal J side. Comparison of the individual waveforms shows that the values obtained during the simulation for both methods coincide, which shows no phase shift, but the voltage values during ground fault duration (from 0.5 to 1 s) for the "b" method are higher, both in the peak and in the rest of the plot. The result of the maximum overvoltages and the conclusions from the obtained values will be presented in Section 4: Discussion.

3.2. Neutral Point Treatment

As shown in Figure 1, the network in which the analyzed cable works is grounded through the T2 grounding transformer (parameters included in Table 1). In the conducted simulations, the variant of operation with the grounding inductor was selected, while in this point, the overvoltages in the network grounded by the inductor or by resistor will be compared. The choice is guided by the actual operation of the MV network and the fact that due to the numerous advantages related to e.g., the operation of protections in the MV network, work with a compensated neutral point is mostly used. However, in the case of the MV network, where there is a larger share of cable lines, it is recommended to use a resistor to ground the neutral point [21,22]. The parameters of the grounding devices have been matched to the other network parameters. The cable line is arranged in a triangle, its diameter is 150/50 and the length is 1 km. The discussed in this section and presented on Figure 5 waveforms are seen from the overhead line in Terminal J.

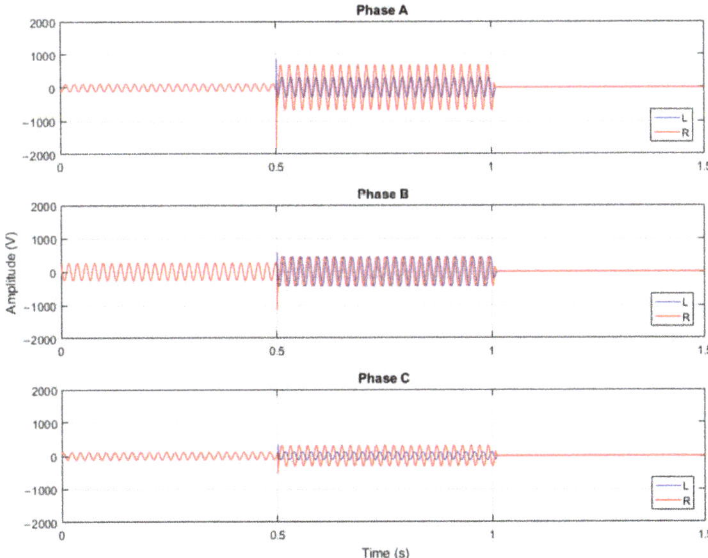

Figure 5. Voltage waveforms in three-phase cable line screens in Terminal J for neutral points grounded by the inductor and resistor.

The presented waveforms show the voltages induced in the cable screens during the ground fault in phase A of the overhead line. As it is shown, the grounding method has quite a significant influence on the voltage waveform, both its value and phase shift. When the neutral point is grounded by a resistor, the ground fault overvoltage is slightly higher in the peak, while the voltage during the duration of the fault, especially for the ungrounded cable screen of phase A, is significantly higher.

3.3. Cable Arrangement

MV cable lines could work in two arrangement types—flat and triangular, as shown in Figure 6. Depending on the system in which they work, the induced values in the conductive elements may change due to the influence of individual phases on each other [5]. The most commonly used system in practice is the flat one, but as a result of possible shifts and movements to the ground, in which the cables are embedded, these configurations may change. A cable arrangement in the three-phase cable line is shown in Figure 6. At this point, possible differences in the voltage induced in the cable screens during the ground fault were analyzed, assuming the basic parameters of the system. The cable line has a diameter of 150/50 and length of 1 km. In the example, phase A of the cable screen remains one-sidedly ungrounded, and in the same phase, a ground fault occurs in the overhead line.

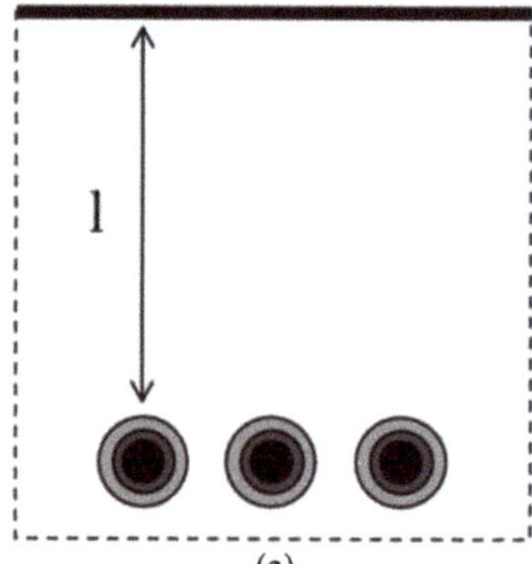

(a)

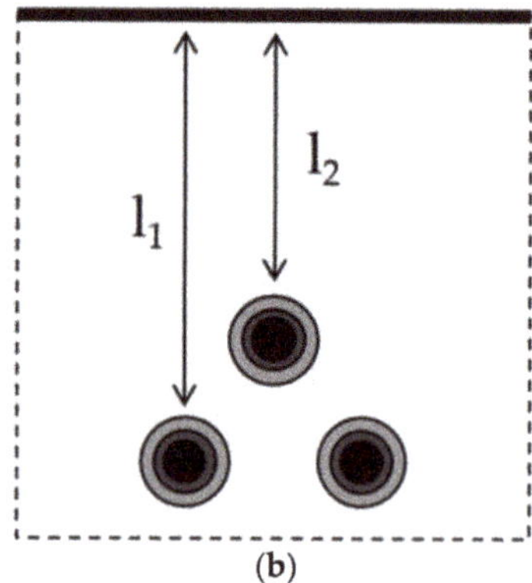

(b)

Figure 6. Cable arrangement in a three-phase cable line: (**a**) flat, where l = 0.7 m is the cable recess, (**b**) triangular, where l_1 = 0.7 m and l_2 = 0.65 m are the cable recesses.

Simulation tests have shown that the change in the voltage level is insignificant from the point of view of overvoltages, and the peak voltage values in ungrounded phase A at the moment of the ground fault occurrence are equal for the flat and triangular arrangement and are 1032 V.

3.4. Conductor Cross-Section

One of the parameters that may affect the level of overvoltages in a cable line is the cross-section of the conductor. For the variant, XRUHAKXS cables with a cable screen of 50 mm^2 cross-section and 120, 150, 185, 240 and 300 mm^2 conductors were adopted. The cable parameters were modeled according to catalog data [15]. The network parameters are equivalent to those from Section 3.1, and the configuration of the system assumes the one-sidedly ungrounded cable screen phase. An additional variable in this part of the simulation is the location of the ground fault. Figure 7 shows the maximum values of overvoltages occurring in the cable line phases during a ground fault in phase A of the overhead line, while Figure 8 shows maximum values of overvoltages. When the ground fault occurs in phase B, the cable screen of phase A remains ungrounded.

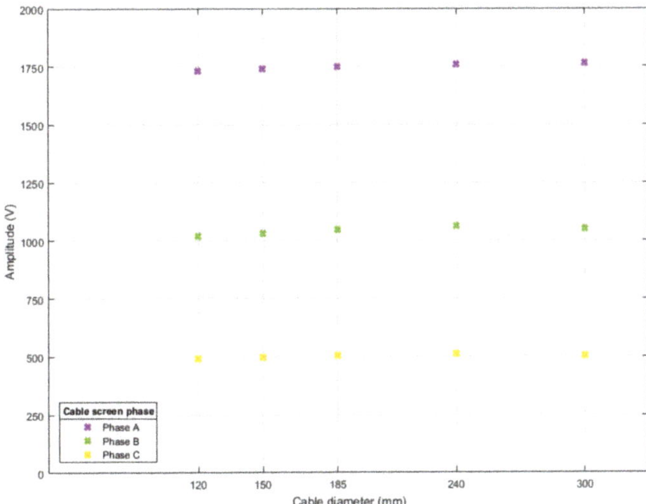

Figure 7. Maximum overvoltage in Terminal J for different conductor cross-sections during a ground fault in phase A.

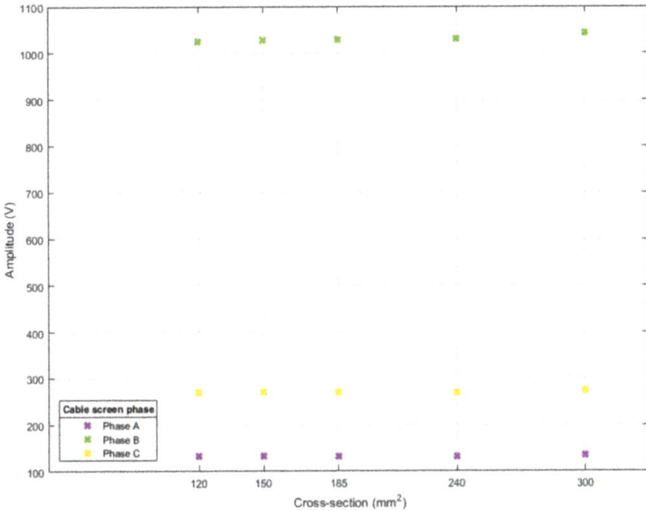

Figure 8. Maximum overvoltage in Terminal J for different conductor cross-sections during a ground fault in phase B.

The overvoltage values obtained in the computer simulations appearing in the ungrounded cable screen and in the grounded cable screen show the relationship between the overvoltage level and the cross-section of the conductor in the cable line. The comparison of the values presented in Figures 7 and 8 also shows a strong relationship between the overvoltage in the cable screen and the phase in which the ground fault current flows. For the considered example with a fault in phase A, it can be seen that it is in the grounded phase where the highest level of overvoltage occurs, while in case of a fault in phase B, in ungrounded phase A, the overvoltages are the smallest. For all the cores, the maximum overvoltage in the cable screen during the occurrence of the ground fault in phase A oscillates around 1750 V for the phase A, and during the ground fault in phase B, it oscillates around 130 V. However, it can be seen that the voltage changes slightly with increasing the cross-section of the conductor. Maximum values of overvoltages for other phases are included in Table 5.

3.5. Cable Screen Cross-Section

Another variable for simulation is the cross-section of the cable screen of the XRUHAKXS cable with a conductor cross-section equal to 150 mm^2. The other parameters correspond to those selected in Section 3.4. The two most common cross-sections in practice are 25 mm^2 and 50 mm^2. The computer simulation carried out shows that the comparison of overvoltage waveforms in the form of a figure is difficult due to the overlapping waveforms and voltage values for phase A at the level of $U_{max} = 1744$ for the cable screen cross-section of 25 mm^2 and $U_{max} = 1744$ for the cross-section of 50 mm^2. Overvoltage maximum values for other phases are included in Table 6.

3.6. Cable Line Length

The last variable selected for simulation was cable length. The tests were carried out for a flat cable arrangement with the cross-section of the conductors of the cable as in Section 3.4 of this manuscript and the cross-section of the cable screen equal to 50 mm^2 for each of the considered cables. The ground fault was simulated in phase A of the overhead line, and the work of the cable screen assumes ungrounding its phase A. The length of the cable varies from 0.5 km to 5 km every 0.5 km. During the simulation, the system operated with the neutral point grounded by the reactor. In connection with the results of the maximum ground fault overvoltages or voltage waveforms in the cable screens presented above in Sections 3.1–3.5, only voltages in the cable screen of phase A will be presented and analyzed in this section. This is mainly due to its garment and thus the highest exposure to the high levels of overvoltage among the three phases. Another reason why phase A was selected for analysis at this point is that for the cable screens of the other two phases, the overvoltage does not present such a risk, as they are grounded on both sides. The maximum values of overvoltages for the described cases are presented in Figure 9.

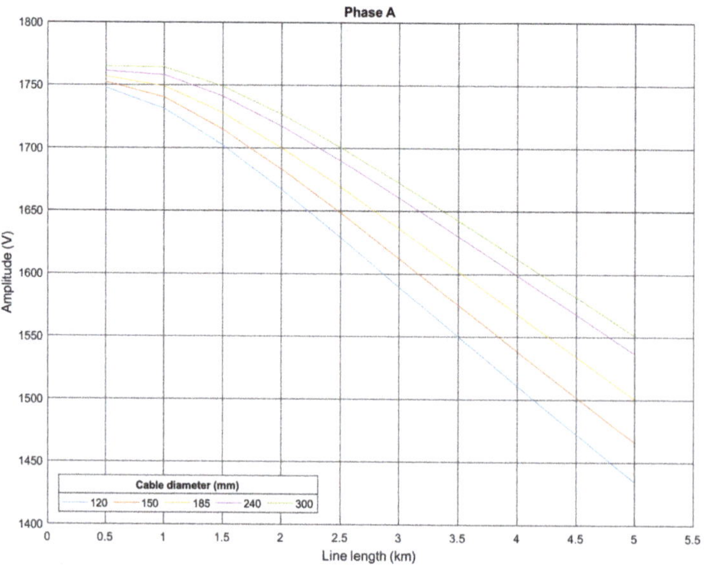

Figure 9. Maximum overvoltage in Terminal J for different conductor cross-sections and lengths in phase A.

The diagram presented above shows that the maximum value of the overvoltage for short cable lines up to 5 km long decreases with the increase of the cable line length, and for the range used for the simulation, it ranges from about 1400 V to about 1800 V.

4. Discussion

The computer simulation presented in Section 3 of the manuscript, which was performed with the use of the MV line model in the DIgSilent PowerFactory program, shows the relationship between the ground fault overvoltages and the basic parameters of the power grid. The presented cases were selected taking into account the current methods of designing and operating the MV network. In this section, tables with summary results of the maximum ground fault overvoltages appearing in the cable screens are presented. This form of presentation of the results may facilitate the assessment of individual options. The assessment will be based on the size criterion. This criterion is based on the electrical strength of the cable screen and assumes that the voltage in it should not exceed 4 kV [23].

The first simulations concerned the configuration of cable screen connections and showed the shaping of overvoltages in terminals to which the cable is connected. In accordance with the results obtained for individual terminals, it was found that the highest overvoltage values appear in Terminal J, in which the cable screen is grounded, and therefore, further overvoltage simulations were performed at this point. Moreover, based on the results of this simulation, which are presented in Table 2, only the configuration with a one-sidedly ungrounded cable screen of phase A (Figure 2b) was taken into account for further considerations. For the configuration with ungrounded cable screens of phases A and B, the overvoltage values, especially in phase A where the ground fault occurred simultaneously, are much greater than the value adopted as the evaluation criterion.

Table 2. Maximum overvoltage for cable line configuration.

	Terminal I		Terminal J	
Phase	**A**	**B**	**A**	**B**
	U_{max} **[V]**	U_{max} **[V]**	U_{max} **[V]**	U_{max} **[V]**
Phase A	101.8	113.2	1740	5998
Phase B	50.67	55.57	1032	2703
Phase C	54.52	116.8	498.7	1431

The next research looked at the effect of how the neutral point works. The results for these simulations are presented in Table 3. With the ungrounded cable screen of phase A, the overvoltage values did not exceed 2 kV, but it is clearly visible that for the network with a neutral point grounded through a resistor, the overvoltage values are higher. This is a very important observation, because when designing cable networks, it is recommended to ground the neutral point of the network through a resistor [22].

Table 3. Maximum overvoltage for different neutral point grounding devices.

	Terminal J	
Phase	**Neutral Point Grounded by Resistor**	**Neutral Point Grounded by Reactor**
	U_{max} **[V]**	U_{max} **[V]**
Phase A	1819	1740
Phase B	1145	1032
Phase C	520.4	498.7

Subsequent tests concerned parameters related to the cable's construction and method of its laying. Table 4 shows the maximum overvoltages for the flat and triangular systems. As can be seen, these values for the cable screen in any phase, regardless of grounding configuration, do not exceed the size criterion, but they also do not change significantly, which leads to the conclusion that the cable arrangement will not affect the overvoltage value.

Table 4. Maximum overvoltage according to the cable system layout.

Phase	Terminal J	
	Flat Layout	Triangular Layout
	U_{max} [V]	U_{max} [V]
Phase A	1740	1740
Phase B	1032	1023
Phase C	498.7	498.3

Tables 5 and 6 summarize the results of the maximum overvoltage values for the systems with different cross-sections for the conductors but also with different phases in which ground fault occurs. As can be seen from the example with the ungrounded cable screen of phase A, when the ground fault occurs in an ungrounded phase (Table 5), the overvoltage values are apparently the highest for this phase. In the case of the ungrounded cable screen in phase A, but the ground fault in phase B (Table 6), the overvoltage value is insignificant. On the basis of these results, it can be also concluded that the cross-section of a working conductor only slightly influences the maximum overvoltage. Similar conclusions can be drawn by comparing the results in Table 7, which lists the maximum voltages for cables with cross-sections of cable screen of 25 mm^2 and 50 mm^2. Again, the cross-section of this element is not important from the overvoltage point of view.

Table 5. Maximum overvoltages for different cross-sections of conductors.

Phase	Ground Fault in Phase A—Terminal J				
	120/50	150/50	185/50	240/50	300/50
	U_{max} [V]	U_{max} [V]	U_{max} [V]	U_{max} [V]	U_{max} [V]
Phase A	1731	1740	1749	1758	1764
Phase B	1021	1032	1048	1063	1051
Phase C	493.7	498.7	506.2	513.3	505.4

Table 6. Maximum overvoltages for different cross-sections of conductors.

Phase	Ground Fault in Phase B—Terminal J				
	120/50	150/50	185/50	240/50	300/50
	U_{max} [V]	U_{max} [V]	U_{max} [V]	U_{max} [V]	U_{max} [V]
Phase A	133.6	133.5	132.9	132.4	135.5
Phase B	1025	1028	1029	1031	1043
Phase C	270.4	270.7	270.1	269.8	274.4

Table 7. Maximum overvoltages for different cross-sections of cable screens.

Phase	Terminal J	
	150/25	150/50
	U_{max} [V]	U_{max} [V]
Phase A	1744	1740
Phase B	1022	1032
Phase C	491.9	498.7

The last simulations are the output of all previous simulations. For the ungrounded cable screen of phase A, in which the ground fault occurs, with the same cross-section of the the cable screen, a neutral point grounded through the reactor, and different cross-sections of conductors, the influence of the line length on the overvoltage is shown. Based on the results presented in Table 8, it can be concluded that short cable lines are particularly exposed to high overvoltage values.

Table 8. Maximum overvoltages according to length of cable line.

Length [km]	Terminal J, Phase A				
	120/50	150/50	185/50	240/50	300/50
	U_{max} [V]	U_{max} [V]	U_{max} [V]	U_{max} [V]	U_{max} [V]
0.5	1748	1752	1757	1761	1765
1	1731	1740	1749	1758	1764
1.5	1702	1715	1728	1741	1749
2	1667	1683	1700	1717	1727
2.5	1629	1648	1669	1690	1701
3	1589	1612	1636	1660	1672
3.5	1550	1575	1602	1630	1642
4	1511	1538	1568	1599	1612
4.5	1472	1502	1534	1568	1581
5	1434	1466	1500	1537	1551

5. Conclusions

Summarizing the results for the overvoltages in the cable screens obtained as a result of computer simulations and taking into account the dependencies described in the Discussion section, conclusions can be drawn about the influence of the parameters of the variables selected for consideration in the described model. According to the analysis of the phenomena, it can be concluded that the one-sidedly ungrounding of one of the cable screen phases causes the possibility of an overvoltage hazard for the ungrounded cable screen phase, regardless of other conditions in the network. However, these conditions may have an impact on the level of these overvoltages and, knowing these dependencies, it is possible to use this method of cable screen operation. The manuscript adopted the limit values of the ground fault overvoltage equal to 4 kV; however, taking into account, for example, the issues of protection against electric shock and the possibility of transferring potentials, this value can be freely adjusted to the application, and thus, it will be possible to determine the correctness of the cable screen's ungrounding. Therefore, the assumptions presented here must also consider the conditions of protection against electric shock and safe voltage levels in a cable network, where these requirements are much lower than in the case of an overhead line. In general, if one-sidedly ungrounding of the cable screen phase will be used, it is suggested to equipped this phase with a spark-gap surge arrester to ensure safe operation of the system.

In connection with the obtained results, it can be assumed that these are not all the variables that should be considered in the overvoltage study. In the described case, the network structure is very limited and it has minimal load. If the load is increased, the current flowing to the power supply can have a significant effect on the overvoltage that occur in the cable screen. In further works, the dependence of the overvoltage value on the current load of a given cable line should also be considered. It also seems important to find the place where the ground fault occurred and to check what the impact on the overvoltage would be if the ground fault changed its place from overhead line to cable line.

Author Contributions: Conceptualization, K.W.; Formal analysis, A.S.-S. and K.W.; Investigation, A.S.-S.; Methodology, A.S.-S. and K.W.; Software, A.S.-S.; Supervision, K.W.; Validation, A.S.-S.; Visualization, A.S.-S.; Writing—original draft, A.S.-S.; Writing—review and editing, K.W. All authors have read and agreed to the published version of the manuscript.

Funding: This research was financing from Ministry of Science and Higher Education, Poland, grant number 0711/SBAD/4455.

Data Availability Statement: Not applicable.

Conflicts of Interest: The authors declare no conflict of interest.

References

1. *IEEE Guide for Bonding Shields and Sheaths of Single-Conductor Power Cables Rated 5 kV through 500 kV*; IEEE Standard Association: Piscataway, NJ, USA, 2014.
2. Sarajcev, I.; Majstrovic, M.; Goic, R. Determining currents of cable sheaths by means of current load factor and current reduction factor. In Proceedings of the 2003 IEEE Bologna Power Tech Conference Proceedings, Bologna, Italy, 23–26 June 2003; Volume 3, p. 4.
3. Lowczowski, K. Badanie wpływu ułożenia kabli na straty energii w żyle powrotnej–symulacja w programie PowerFactory. *Przegląd Elektrotechniczny* **2016**, *10*, 54–57. [CrossRef]
4. Lowczowski, K.; Lorenc, J.; Zawodniak, J.; Dombek, G. Detection and location of Earth fault in MV feeders using screen earthing current measurements. *Energies* **2020**, *13*, 1293. [CrossRef]
5. Novak, B.; Koller, L.; Berta, I. Loss reduction in cable sheathing. *Renew. Energy Power Qual. J.* **2010**, *1*, 293–297. [CrossRef]
6. Lin, Y.; Xu, Z. Cable sheath loss reduction strategy research based on the coupled line model. *IEEE Trans. Power Deliv.* **2015**, *30*, 2303–2311. [CrossRef]
7. Jiaming, L.; Long, X.; Junbo, D. Analysis of Reclosing Overvoltage in Cable-Overhead Line Hybrid Line Based on Return Circuit Theory in Modal Domain. *Trans. China Electrotech. Soc.* **2020**, *35*, 1754–1763.
8. Czapp, S.; Dobrzynski, K.; Klucznik, J.; Lubosny, Z. Analysis of induced voltages in sheaths of high voltage power cables for their selected configuration. In Proceedings of the XVIII Międzynarodowa Konferencja Aktualne Problemy w Elektroenergetyce APE'17, Jastrzebia Gora, Poland, 7–9 June 2017.
9. Andruszkiewicz, J.; Lorenc, J.; Lowczowski, K.; Weychan, A.; Zawodniak, J. Energy losses' reduction in metallic screens of MV cable power lines and busbar bridges composed of single-core cables. *Eksploat. I Niezawodn. Maint. Reliab.* **2020**, *22*, 15–25. [CrossRef]
10. Heiss, A.; Balzer, G.; Schmitt, O.; Richter, B. Surge arresters for cable sheath preventing power losses in MV networks. In Proceedings of the 16th International Conference and Exhibition on Electricity Distribution, CIRED, Amsterdam, The Netherlands, 18–21 June 2001.
11. Schött-Szymczak, A.; Walczak, K. Analysis of overvoltages appearing in one-sidedly ungrounded MV power cable screen. *Energies* **2020**, *13*, 1821. [CrossRef]
12. Chowdhuri, P. *Electromagnetic Transients in Power System*; John Wiley & Sons: New York, NY, USA, 1996.
13. Lowczowski, K.; Lorenc, J.; Tomczewski, A.; Nadolny, Z.; Zawodniak, J. Monitoring of MV cable screens, cable joints and earthing systems using cable screen current measurements. *Energies* **2020**, *13*, 3438. [CrossRef]
14. Transformer 110/15 Datasheet. Available online: http://www.ftz.pl/download/Transformatory%20Mocy.pdf (accessed on 23 March 2021).
15. Medium Voltage Cables with XLPE Insulation Datasheet. Available online: https://www.nkt.com.pl/fileadmin/user_upload/Products/Data_sheets/NA2XS_F_2Y__XUHAKXS__12_20_kV_DS_PL_EN.pdf (accessed on 23 March 2021).
16. Grounding Transformer Datasheet. Available online: http://www.ftz.pl/transformatory-uziemiajace-olejowe.php (accessed on 23 March 2021).
17. Overhead Line PAS70 Datasheet. Available online: https://www.eltrim.com.pl/wp-content/uploads/2017/03/eltrim_final-1.pdf (accessed on 23 March 2020).
18. Lavrov, Y.A.; Korobeynikov, S.M.; Petrova, N.F. Protection of cable with XLPE insulation from high-frequency overvoltages. In Proceedings of the 2016 11th International Forum on Strategic Technology (IFOST), Novosibirsk, Russia, 1–3 June 2016.
19. DIgSILENT Power Factory Knowledge Base. 2020. Available online: https://www.digsilent.de/en/faq-powerfactory.html (accessed on 23 March 2021).
20. Hoppel, W. *Medium Voltage Networks. Power System Protection and Protection against Electric Shock*; PWN: Warsaw, Poland, 2017.
21. Drandić, A.; Marušić, A.; Drandić, M.; Havelka, J. Power system neutral point grounding. *J. Energy* **2017**, *66*, 52–68.
22. Duda, D.; Szadkowski, M. Surge protection sheath of HV cables in various system of cable screen conections. *Przeglad Elektrotechniczny* **2014**, *90*, 37–40. [CrossRef]
23. British Standards Institution. *Earthing of Power Installations Exceeding 1 kV a.c.*; BSI: London, UK, 2012.

 *energies*

Article

Artificial Negative Polarity Thunderstorm Cell Modeling of Nearby Incomplete Upward Discharges' Influence on Elements of Monitoring Systems for Air Transmission Lines

Nikolay Lysov *, Alexander Temnikov, Leonid Chernensky, Alexander Orlov, Olga Belova, Tatiana Kivshar, Dmitry Kovalev and Vadim Voevodin

Department of Electrophysics and High Voltage Technique, Moscow Power Engineering Institute, National Research University, 111250 Moscow, Russia; TemnikovAG@mpei.ru (A.T.); ChernenskyLL@mpei.ru (L.C.); OrlovAV@mpei.ru (A.O.); belovaos@mail.ru (O.B.); geratk@mail.ru (T.K.); kovalevdi@mpei.ru (D.K.); voevodinvv@mpei.ru (V.V.)
* Correspondence: streamer.corona@gmail.com; Tel.: +7-9168456253

Abstract: The article represents results of a physical simulation of incomplete upward leader discharges induced on air transmission lines' elements, using charged artificial thunderstorm cells of negative polarity. The influence of such discharges on closely located model sensors (both of rod and elongated types) of digital monitoring systems, as well as on the models of receiver-transmission systems of local data collection (antennas), was determined. Effect of heterogeneity of electromagnetic field caused by incomplete upward discharges on frequency specter of signals generated on sensors and antennas was estimated. Wavelet analysis was carried out to determine the basic frequency diapasons of such signals. Based on experimental data obtained, suppositions about the extent of influence of nearby incomplete leader discharges on the functioning of currently used systems of transmission lines' monitoring were made.

Keywords: artificial thunderstorm cell; lightning; upward leader discharges; electromagnetic radiation spectrum; wavelet; transmission line monitoring system; model element; simulation

Citation: Lysov, N.; Temnikov, A.; Chernensky, L.; Orlov, A.; Belova, O.; Kivshar, T.; Kovalev, D.; Voevodin, V. Artificial Negative Polarity Thunderstorm Cell Modeling of Nearby Incomplete Upward Discharges' Influence on Elements of Monitoring Systems for Air Transmission Lines. *Energies* **2021**, *14*, 7100. https://doi.org/10.3390/en14217100

Academic Editor: Dan Doru Micu

Received: 31 August 2021
Accepted: 26 October 2021
Published: 31 October 2021

Publisher's Note: MDPI stays neutral with regard to jurisdictional claims in published maps and institutional affiliations.

1. Introduction

Remote monitoring systems are increasingly being introduced into the management of transmission lines and become an important element of digital power industry [1–11]. During their exploitation, it was observed that the functionality of the artificial intelligence elements used in such systems (i.e., all devices that contain electromagnetic parts, e.g., sensors; analog–digital converters for processing of recorded signals; antennas, etc.) may be compromised under the influence of lightning and atmospheric electricity [3,11,12]. It is not entirely clear how the discharge phenomena forming on both rod and elongated ground objects under the influence of thunderclouds and lightning (e.g., flashes of streamer corona, ascending leaders, the main discharge) and the electromagnetic radiation they create will affect functioning of such elements [13,14]. The purpose of this work is to estimate the influence of electromagnetic radiation emerged from incomplete upward leader discharge phenomena, located close to the remote monitoring system, on the system's elements. To achieve this goal, a thunderstorm experimental environment was modeled using artificially created thunderstorm cells of negative polarity. This work is a continuation of [15], where the characteristics of streamer discharges that did not pass into the ascending leader were considered.

2. Experimental Complex and Experiment Schemes

Physical processes were modeled on experimental measurement complex "Thunderstorm" [16], as in the scheme of Figure 1.

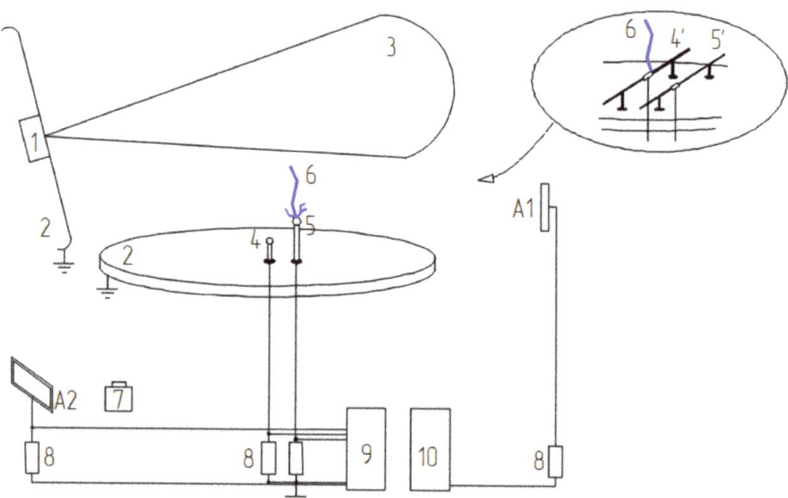

Figure 1. First scheme of experimental and measurement setup: 1—charged aerosol generator, 2—grounded electrostatic screens, 3—artificial thunderstorm cell, 4, 5—rod electrodes, 4′, 5′—elongated electrodes, 6—upward discharge phenomena, 7—digital camera Panasonic DMC-50, 8—shunts, 9, 10—digital memory oscilloscopes Tektronix TDS 3054B иTektronix DPO 7254, A1, A2—flat antennas.

Experimental measurement complex "Thunderstorm" was created at the Department of Electrophysics and High Voltage Technique (National Research University "Moscow Power Engineering Institute") with the purpose of physical modeling and exploration of fundamental and practical problems in the fields of atmospheric electricity, lightning physics, and lightning protection of both grounded and flying objects [17–21]. It allows creating artificial clouds from a highly charged water aerosol with the maximal density of the volume charge (the artificial thunderstorm cells), which are equal to the electrically charged thunderclouds. Due to the capacity to create artificial thunderstorm cells of both polarities, located above each other, the complex allows to generate different types of electric discharges characteristic of thunderclouds, from weak diffuse discharges to the main stage of discharge with high current release.

Experimental investigational complex "Thunderstorm" consists of an aerosol camera (volume > 120 m^3), generator of charged aerosol, investigational object (the studied system of electrodes), and measuring complex. In the aerosol camera, two varieties of artificial aerosol clouds (artificial thunderstorm cells) were used: either a one-cell structure of negative or positive polarity, or two-cell structure consisting of two artificial thunderstorm cells of the same or different polarities, located above ground at different heights [22,23].

A condensational generator of charged water aerosol was used to create artificial thunderstorm cells. When the pressure in the generator was between 4 and 10 bar, the velocity of water steam efflux from subsonic convergent nozzle was around 400 m/s. Water aerosol particles were charged by condensation on ions and ion charge in the field of corona discharge occurring in the nozzle exit section.

The charged aerosol generator maintained an outlet current in a broad diapason from 10 μA to 150 μA. As a result of variations in the generator's outlet current, the charge of an individual artificial thunder cell could be regulated in a broad value diapason, reaching up to several hundreds of microcoulombs (at maximal charging currents). Volume density of charge in the central area of the cloud was in the diapason of 1.5×10^{-4}–1.5×10^{-2} C/m^3. Each of the charged aerosol clouds is several cubic meters in volume. The lower cell is located approximately at the 1 m height over the grounded plane, while the higher cell is located at the 2.1 m height.

When the generator is working at a high outlet current, the maximum potential of a single artificial thunderstorm cell may reach up to 1.0–1.5 MB [24–27]. As a result, a strong electric field is formed in the area between the artificial thunderstorm cell and the grounded plane. Its intensity reaches up to 6–10 kV/cm close to the ground and up to 17–20 kV/cm at the cell's lower border. Moreover, the electric field of the almost quasi-homogeneous quality is formed in the area between the cell's lower border and the ground plane at the height up to 0.4 cm from the ground, and at a distance of 1.0 to 2.0 m from the generator's nozzle [24,27]. The value of electric field's intensity in this area changes comparatively slowly as the height grows (30–40% increase). Due to this, investigative objects located in this area (e.g., rod or other electrodes up to 0.4 m in height) will be similarly affected by the electric field of the charged aerosol cloud.

The dynamics of electric field formation in the area between the artificial thunderstorm cell and the ground shows that 50 ms after the beginning of the water aerosol flow charging, field intensity at the ground under the charging cell increases to 1 kV/cm, and one second after, it may be more than 5–6 kV/cm [25–27]. By this time, the electric field intensity at the lower border of the artificial thunderstorm cell's charged areas is 10–12 kV/cm. These conditions facilitate discharge formation on the model investigative objects located on the grounded plane.

Therefore, a slow increase in the electric field's intensity is observed in the significant area of the space between an artificial thunderstorm cell and ground. Due to the inertial character of the charged cloud formation and existence of negative feedback between the cell and occurring discharge phenomena, it can thus be expected that all discharge processes in the space between the cell and grounded model object will occur and develop under minimal required conditions practically without overvoltage. In this case, the character of distribution and value of electric field's intensity between an artificial thunderstorm cell and the grounded plane are comparable to characteristics of the field under a real thundercloud. Such a field is characteristic of the conditions when a thundercloud (or a downward lightning leader beginning to develop from the cloud) affects all above-ground objects [28]. By changing the generator's outlet current, it is possible to emulate the development of all types of sparks characteristic for a thunderstorm, including leader stage and main discharge.

Figures 2–4 demonstrate photos of the different discharge phenomena, forming either between cells of different polarity or between cells and grounded objects. In recent years, a similar complex based on a generator of charged aerosol has been used in Istra (Moscow region) (Rakov V.A., Syssoev V.S., Kostinskiy A.Y. et al. [29–34]). This generator was also developed at the Department of Electrophysics and High Voltage Technique (National Research University "Moscow Power Engineering Institute").

More than 90% of the lightning strikes into the ground are negative. This is related to the characteristic thundercloud structure, where, in the majority of cases, the main negative charge is located in its lower half [28,35]. Its impact leads not only to initiation and formation of the downward lightning leader, but also to initiation and formation of the upward discharges, including incomplete and/or counterpart positive streamer and leader discharges in the opposite direction from the structural elements of objects located on the ground under the thundercloud [35]. This is why an artificial thunder cell of negative polarity was used for the physical modelling of the possible effect of incomplete upward streamer and leader discharges on the nearby located various elements of power lines' monitoring systems. At the same time, the artificial thunderstorm cell charge (outlet current of aerosol generator) was maintained at a level where the probability of upward discharges' transition into a high-powered main stage was minimal.

Figure 2. Characteristic picture of the discharge formation between artificial thunderstorm cell of negative polarity and the ground.

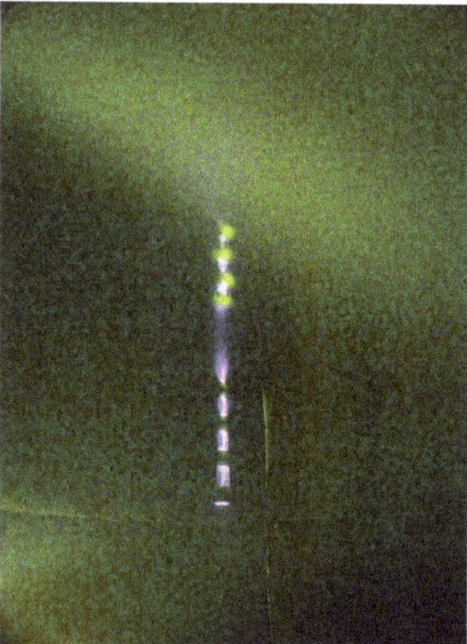

Figure 3. Characteristic picture of the discharge formation between positively and negatively charged artificial thunderstorm cells, using model hydrometeor groups.

Figure 4. Characteristic picture of the discharge formation between the system of artificial thunderstorm cells of negative polarity and the ground.

The potential of the artificial thunderstorm cell of negative polarity reached up to 1.2 MV. Monitoring system elements and grounded parts of air transmission lines were modeled using rod electrodes with spherical tops or appeared as a cylindrical element isolated from the main part of the whole construction (Figure 1). When discharge phenomena appeared at one of the electrodes, its parameters were registered, as well as the aimed signals at the nearest electrode.

Two experimental series were conducted while modeling the effect of close incomplete upward discharges on the elements of power lines' monitoring systems using artificial thunderstorm cells. In the first series, an experimental scheme with two rod model electrodes with different amplification coefficients (AQ) was used. This series has modeled signals induced on rod sensor models (receiver–transmitter devices) by the electromagnetic radiation of the upward leader discharges forming from the nearby rod elements. In the second series, a similar experimental scheme was used, substituting rod elements with elongated (cylindric) models.

Tops of rod electrodes were electrically isolated from their bodies in order to decrease the influence of the electric induction current on the electrical characteristics of the upward leader discharges forming from the top of the grounded model object, or to extract the signals induced by the nearby upward leader discharge specifically on the rod model sensor. The middle part of the elongated model conductors was isolated from the other parts with the same purpose.

Radius of the spheric (or conic, in case of small radii of curve slope) top of model rod electrodes varied in diapason from 0.3 cm to 2.5 cm. Rod electrode height varied from 16 cm to 40 cm. Radii of metal tubes modelling cylindric sensor elements, phase wires, and ground wires varied from 0.5 cm to 2.5 cm. Elongated electrodes were located at the height between 15 and 37 cm. AQ for model electrodes was calculated according to specialized software BETAFields.

Electrical characteristics of discharge phenomena between artificial thunderstorm cell and grounded model electrodes were registered based on measuring the following parameters:

(1)   Discharge current initiated at the ground model elements,

(2)   Electric induction current inflicted by discharge phenomena on the nearest models of digital objects and systems (sensors and antennas).

Both types of currents were measured using special low-inductance shunts.

During the experiment, the radii of the vertices of rod model objects and the radii of model cylindrical objects varied. This was done in order to model the situation when, under the influence of atmospheric electricity, discharge phenomena form on the digital elements and systems, which have substantially different electric field AQs. During statistical analysis the difference between the objects was taken into account by dividing them into groups according to this coefficient: group I—AQ < 12; group II—12 < AQ < 27; and group III—AQ > 27.

The following variants of discharge phenomena formation next to digital objects were modeled in this work:

-    Streamer corona and upward leader discharge form under the influence of a thundercloud's electromagnetic field on a grounded object (e.g., transmission tower). Electromagnetic radiation from the upward discharge affects the functioning of the sensors and antennas located on or close to the tower (Figure 5a),
-    Streamer corona and upward leader discharge form under the influence of a thundercloud's electromagnetic field on a middle segment of a lightning protection wire. This affects the functioning of the sensors and antennas located in the middle segment of the phase wire (Figure 5b).

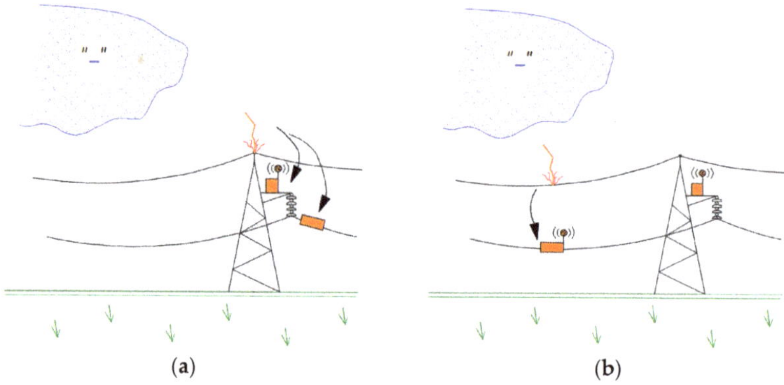

(a)   (b)

Figure 5. Variants of formation of discharge phenomena affecting power lines' digital monitoring systems: (**a**) streamer corona/upward leader discharge forms on a transmission tower, (**b**) streamer corona/upward leader discharge forms on a middle segment of a lightning protection wire.

Figures 6 and 7 represent photos of discharge phenomena that developed on the rod and elongated model electrodes while using the "Thunderstorm" experimental complex. It is clear that the discharge develops only on one of the two model electrodes. Photos were taken on a digital camera in long-exposure mode, which was set to turn on at the same time as the artificial thunderstorm cell's charging process began. This allowed documenting all development stages of the discharge phenomena generated on every element in the experimental area.

Figure 8 represents a characteristic oscillograph chart for the current of streamer flash and incomplete upward leader discharge and signals induced on the nearby grounded element and antennas. The following parameters were registered for the current signal: current amplitude, maximal velocity of current build-up, impulse charge, and total impulse length.

Figure 6. Upward leader discharges from the grounded rod model elements.

Figure 7. Upward leader discharges from the grounded cylinder model element.

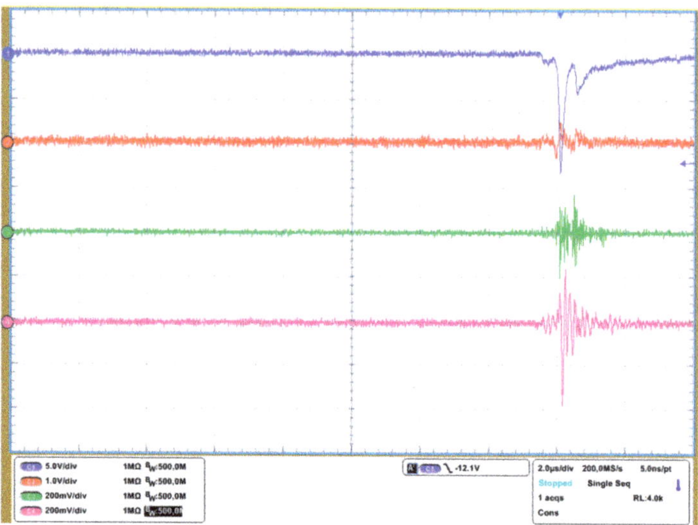

Figure 8. Characteristic oscillograph charts for discharge current from grounded electrode (channel 1), and electric induction currents induced on nearby electrode (channel 2) and flat antennas (channels 3 and 4).

Spectral characteristics of the induced signal were determined using software created specifically for this purpose. It was based on the Mexican hat wavelet analysis [36–38]. Upper level of the characteristic frequency, maximal intensity, and frequency of the maximal intensity in the wavelet spectrum have been determined. Characteristic wavelet spectrums for the currents and induced signal presented in Figure 9 are shown in Figures 10 and 11, correspondingly. These signals correlate to the signals from channels 1 and 2, represented in Figure 8.

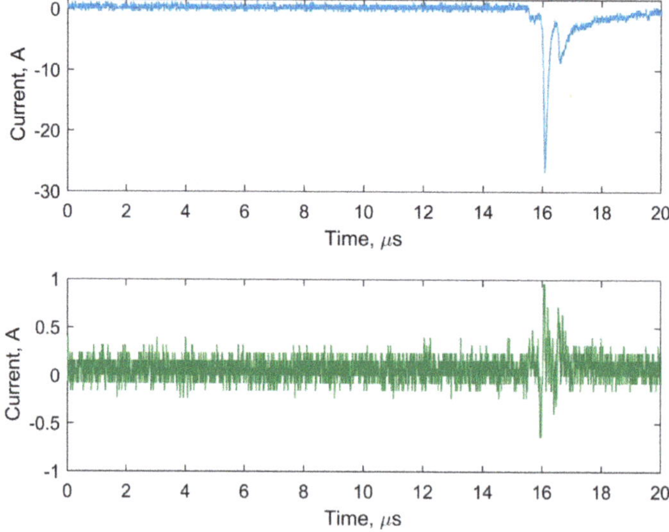

Figure 9. Oscillograms of the upward discharge current (**upper**) and induced electromagnetic effects (**bottom**).

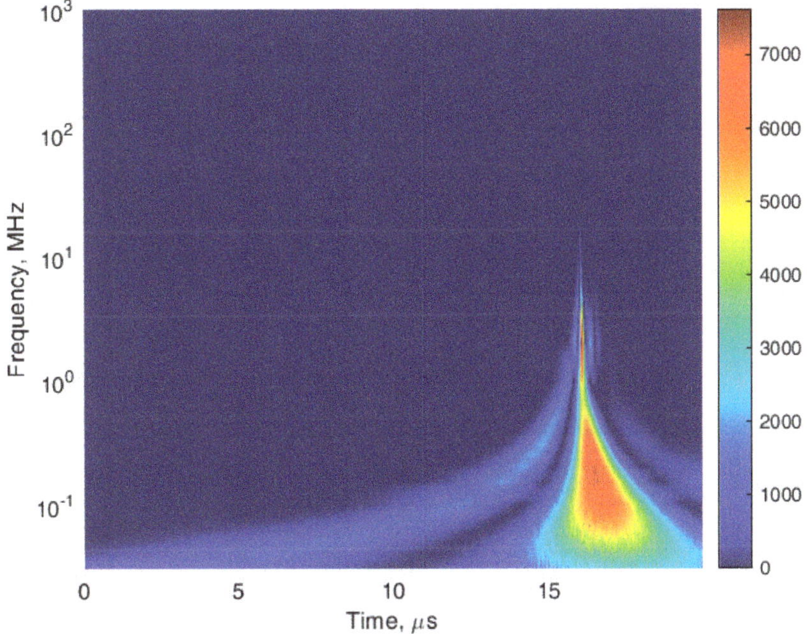

Figure 10. Wavelet spectrum of upward discharge current.

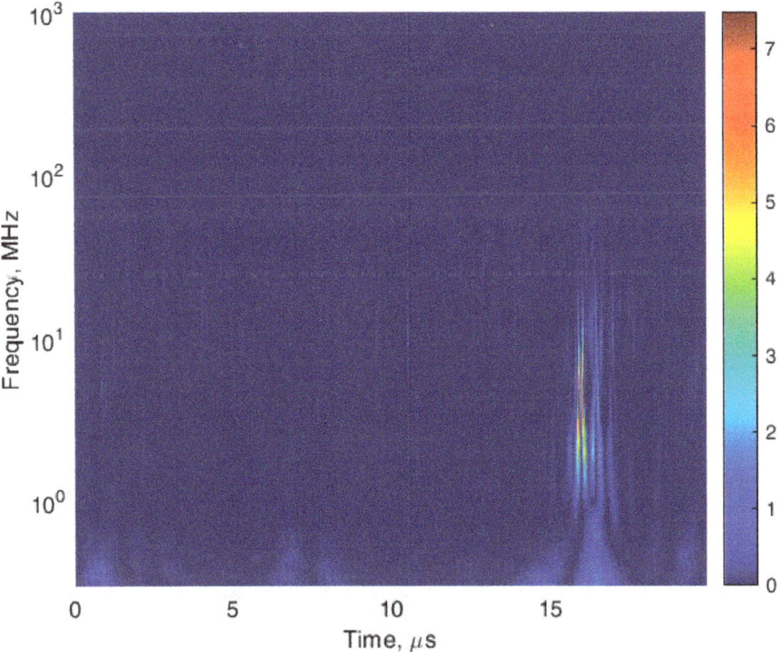

Figure 11. Wavelet spectrum of the signal induced on the nearby model element by the upward discharge.

3. Results

More than 300 experiments were completed and analyzed for this work. The parameters of electrode systems and their location in relation to each other varied through the experiments. Tables 1 and 2 represent results of processed signals and their wavelet analysis.

Table 1. Parameters of signals induced on model rod and elongated elements by a closely located flash of upward discharge (average values).

Source of Effect	Upward Discharge (Rod)		Upward Discharge (Elongated)	
Amplification Coefficient	ΔT, µs	|Imax|, A	ΔT, µs	|Imax|, A
Group I	1.2	0.9	1.1	1.2
Group II	0.3	1.6	0.5	1.3
Group III	0.7	1.0	1.0	1.2

Table 2. Parameters of signals induced on flat antennas with upward discharge (average values).

Antenna	A1		A2	
Parameters	ΔT, ms	|Imax|, A	ΔT, µs	|Imax|, A
Average	1.1	1.1	1.8	1.8
Maximum	4.5	3.1	9.8	3.3

Analysis of spectral parameters of the signals induced on rod and elongated model elements, as incomplete upward discharges form on the nearby located model electrodes, has shown that the surrounding electrical field is considerably influenced by those discharges. Table 3 shows analysis results divided into three groups, depending on the value of electric field's amplification coefficient (AQ).

Table 3. Spectral parameters of signals induced with upward discharge (average values).

Source of Effect	Rod Elements			Elongated Elements			Antennas	
AQ	Group I	Group II	Group III	Group I	Group II	Group III	A1	A2
fmax, MHz	122	756	391	119	785	467	602.3	144.4
f(Cmax), MHz	11.1	77	14.7	9.6	73	12.3	51.6	10.5

During the experiments, several variants of elements' location in relation to each other were modeled. Receiver–transmitter devices (or other sensors) were located either very close to the place of discharge formation or at a longer distance from it (antenna A2). Another location variant had antennas located higher than the place of discharge formation (antenna A1 was located across from the lower cloud border of the artificial thunderstorm cell). This emulates, for example, natural conditions of a hilly or mountain scenery, high-rise objects, or the use of flying machines.

It is clear from the presented results that all variants of model rod and elongated elements of an intelligent monitoring system will be affected by nearby forming discharges, and that the induced signals will include frequency diapasons in hundreds of MHz up to GHz (Figure 12). When a flash of impulse streamer corona forms nearby, the values of maximal frequencies in the wavelet specter will be observed for model sensors with "average" amplification coefficients (group 2, 12 < AQ < 27): 756 MHz for rod elements and 785 MHz for elongated elements, on average. At the same time, on the model sensors with relatively low AQ (group 1, AQ < 12) in most cases, maximal frequencies in the wavelet specter will not surpass 100 MHz (Figure 12). Unlike the cases when the streamer flash did not pass into the ascending leader [15], there are no frequencies more than

1 GHz in the specter of signals induced by the upward discharge (streamer flash plus the upward leader).

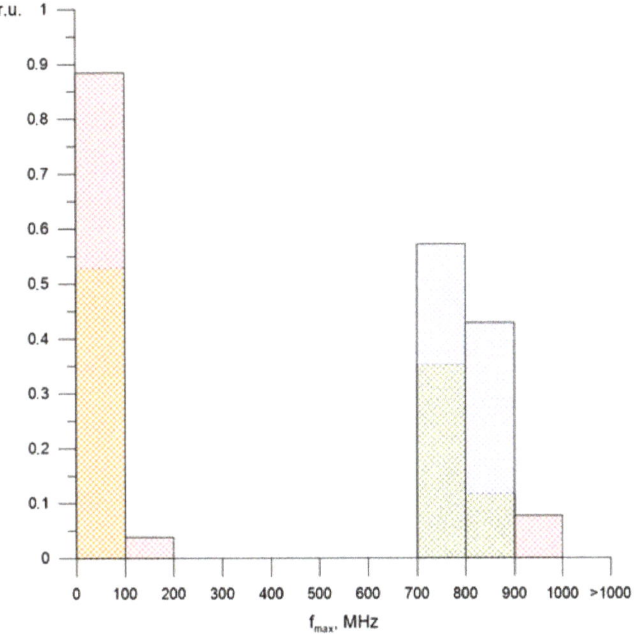

Figure 12. Histogram of diapasons of characteristic maximal frequencies in wavelet specter of the signals induced by close upward discharges on model elements of the rod and elongated types. Group 1—red; Group 2—blue; Group 3—yellow.

AN analogous tendency was discovered for the frequency corresponding to the maximum intensity of Cmax (f(Cmax)) in the wavelet specter (Figure 13): for the model elements in group 2, it was 77 MHz for rod elements and 73 MHz for elongates ones.

Wavelet specter analysis of signal, induced by closely located discharges on flat antennas, which model receiver–transmitter devices and other digital elements, has shown that frequencies of hundreds of MHz up to GHz are present in their specters, even during the primary part of upward discharge formation. Moreover, in the wavelet specter of signals induced on antennas by a close discharge formation, the frequencies f(Cmax) in the wavelet specter will average tens of MHz (Figures 14 and 15).

For avalanches that form and develop on electrodes of various sizes in electrical fields, their electromagnetic radiation (according to [39–44]) may be explicitly expressed in the diapasons from 4.0–5.0 MHz to 0.9–1.0 GHz. Characteristic frequency diapasons of wavelet specter signals (induced on model rod and elongated elements by the nearby discharges) correlate with this mechanism of electromagnetic radiation formation.

As shown by modelling [40–42], the specter of electromagnetic radiation of critical avalanche (which allows for its transition into a streamer) depends on the value of electric field intensity and on the particularities of the change in such field in the area where avalanches are formed. Probably, this is the reason for the very high frequencies detected in the wavelet specter of the signal induced on the model electrode by electromagnetic radiation of the nearby avalanche-streamer system of the upward discharge (impulse streamer corona plus the upward leader), forming on the electrodes with high AQ values (groups 2 and 3). Upward discharges forming on model elements with lower AQs (group 1), in the vicinity of which variation of field intensity occurs lass rapidly, will induce the signals with lower frequency diapasons in their specters.

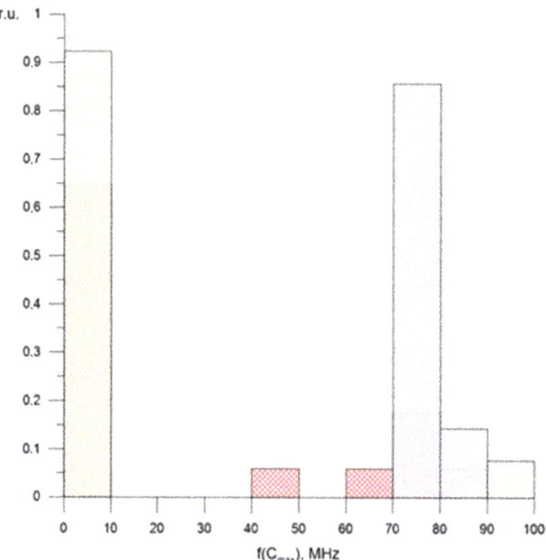

Figure 13. Histogram of f(Cmax) (frequency diapason corresponding to the maximum intensity of Cmax) in the wavelet specter of signals induced by the nearby located upward discharges on model elements of the rod and elongated types. Group 1—red; Group 2—blue; Group 3—yellow.

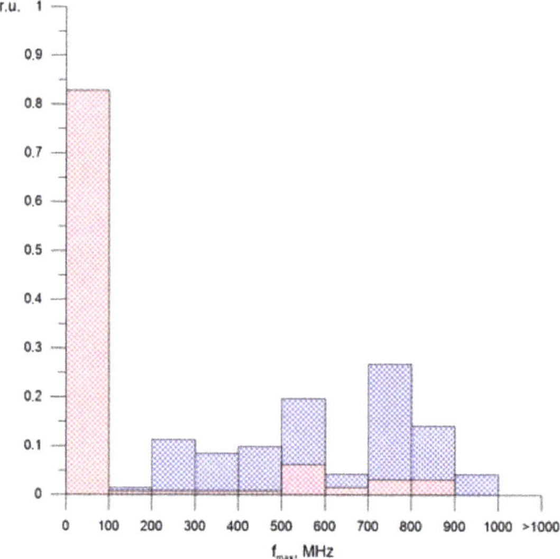

Figure 14. Histogram of diapasons of characteristic maximum frequencies in wavelet specter of signals induced on the flat antennas by a nearby discharge: antenna A1 (in blue); antenna A2 (in red).

It is also important to consider that, since the streamers form and develop by avalanches induced from their heads in electric fields of various size, electromagnetic radiation of those avalanches (according to [39–44]) may be explicitly expressed in diapasons from 0.5–1.0 MHz to 8–10 MHz.

During the formation of a specter of electromagnetic influence of the upward discharge developing between artificial thunderstorm cells and ground, the specter's ultrahigh frequency can also be affected by the powerful streamers of the upward leader reaching to the cell borders. External electric field intensity next to these streamers rose up to 18–22 kV/cm. In such event, several avalanche-streamer strings will be located in this area at the same time. As noted in [39–44], in this case avalanches are situated close in time with each other, and their electromagnetic radiation fields almost overlap, which should lead to an increase in spectral amplitude.

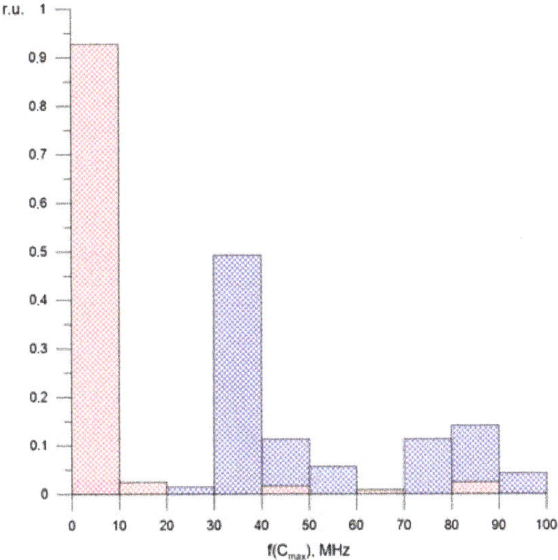

Figure 15. Histogram of diapasons of f(Cmax) in wavelet specter of signals induced on the flat antennas by a nearby discharge: antenna A1 (in blue); antenna A2 (in red).

4. Discussion

Analysis of experimental data has shown the following characteristic spectral diapasons of signals induced on model rod objects of different AQ groups by the nearby streamer corona:

Group 1 objects:

Maximum frequencies (fmax) in the wavelet specter of induced signals: 11–170 MHz; 900–1000 MHz. Frequencies corresponding to the maximal intensity of the wavelet specter (f(Cmax)) of induced signals: 0.1–8 MHz; 90–100 MHz.

Group 2 objects:

fmax: 765–833 MHz. f(Cmax): 73–88 MHz.

Group 3 objects:

fmax: 13–95 MHz; 717–845 MHz. f(Cmax): 0.9–7 MHz; 40–50 MHz, 63–88 MHz.

Analysis of experimental data for model elongated elements (objects) of different AQ groups has shown:

Group 1 objects:

fmax: 5–70 MHz; 900–1000 MHz. f(Cmax): 0.7–9 MHz; 90–100 MHz.

Group 2 objects:

fmax: 711–805 MHz. f(Cmax): 70–82 MHz.

Group 3 objects:

fmax: 4–75 MHz; 735–880 MHz. f(Cmax): 0.3–6 MHz; 40–50 MHz.

In presence of the electromagnetic field generated by a thundercloud and lightning, incomplete upward discharges may form on different grounded objects. Electromagnetic radiation of such discharges, when they develop close to sensors of digital objects and systems, can create risks for the energetic safety of this equipment. The frequencies present in the specter of the signal induced by the nearby streamer flash and upward leader on the sensors may affect their functioning and lead to errors in the collected data.

Such a problem is relevant for the sensors with frequency diapason under 100 kHz and under 1.5 GHz, which control isolation status of high voltage equipment and electromagnetic levels on substation switchyards. Such frequencies are typical for model elements (sensors) of all amplification coefficient groups [13,14]. Analogous errors may occur when electromagnetic radiation from the streamer corona flash and upward leader affects the type of system sensors which employs analog–digital converters (ADC) with working frequencies from several hundred kHz to several GHz in order to measure and further digitally process their signals [12,14]. Additionally, the aforementioned specters of signals induced by the streamer corona flashes transformed into upward leaders on rod sensors of monitoring, diagnostics, and management systems and may cause false reactions and errors for the systems that employ ADCs with working frequencies from several tens/hundreds Hz in slow high-voltage ADCs to several hundred MHz in high-velocity high-performance ADCs [12].

Electromagnetic radiation of incomplete upward discharges may lead to errors in monitoring systems using elongated sensors working in frequency diapasons under 1000 MHz (this relates to sensors belonging to all AQ groups) [3,13,14]. Using analog–digital converters with working frequencies from several hundred kHz to 1 GHz to process data from the signals may also lead to failures [12,14]. Additionally, aforementioned specters of signals induced by the streamer corona and upward leader discharges on elongated and rod sensors of monitoring systems may cause false reactions and errors for the systems that employ ADCs with working frequencies from several tens/hundreds Hz in slow high-voltage ADCs to several hundred MHz in high-velocity, high-performance ADCs [12].

Analysis of experimental data has shown the following characteristic spectral diapasons of signals induced on model flat broadband antennas by the nearby upward discharges:

Antenna A1:

fmax: 111–955 MHz.

f(Cmax): 0.3–6 MHz; 28–60 MHz; 73–98 MHz.

Antenna A2:

fmax: 21–852 MHz.

f(Cmax): 5–15 MHz; 48–88 MHz.

The electromagnetic radiation of incomplete upward discharges described above may also substantively affect receiver–transmitter devices (antennas) of digital systems and objects, which, in turn, may create risks for their energetic safety. Frequencies present in the specter of the signal induced on antennas by a nearby streamer flash and upward leader may lead to errors in their functioning, as well as distortion or loss of transmitted information [1,12]. Firstly, it may affect the data transmission channels using GSM (900/1100 MHz), as well as other systems of data transmission/reception working in MHz diapason. These may include:

- Transmission systems working from sensors with frequency of 915 MHz, located on air transmission lines [1];
- Systems, transmitting data on relatively small distances, e.g., using frequencies 434 MHz, 868 MHz for autonomous monitoring systems;
- Satellite connection systems, e.g., using user-accessible frequency diapason: Orbcomm system with frequencies of transmitters 137.0–138.0 MHz, 400.05–400.15 MHz, frequencies of terminals 148.0–150.05 MHz; "Gonets" system services its users in the diapasons of 0.2–0.3 GHz and 0.3–0.4 GHz [45,46].

5. Conclusions

Completed analysis of experimental data has allowed identifying the influence of close incomplete upward discharges (streamer flash plus the upward leader) on systems of monitoring, diagnostics, and management of the power objects. The influence of electromagnetic radiation of such discharges induced on the digital systems' model sensors and antennas (of both rod and elongated types) has been estimated. Characteristic diapasons of working frequencies of receiver–transmitter systems and digital–analog converters, where such influence is most substantial and may increase risks of failures and errors in their functioning, have been determined.

Author Contributions: Project administration, methodology A.T.; software, L.C. and O.B.; visualization, T.K.; investigation, A.O. and N.L.; writing—original draft preparation, V.V.; funding ac-quisition, D.K. All authors have read and agreed to the published version of the manuscript.

Funding: This study conducted by Moscow Power Engineering Institute was financially supported by the Ministry of Science and Higher Education of the Russian Federation (project No. FSWF-2020-0019).

Institutional Review Board Statement: Not applicable.

Informed Consent Statement: Not applicable.

Data Availability Statement: Not applicable.

Conflicts of Interest: The authors declare no conflict of interest.

References

1. McCall, J.C.; Spillane, P.; Lindsey, K. Determining Crossing Conductor Clearance Using Line-Mounted LiDAR. In Proceedings of the CIGRE US National Committee 2015 Grid of the Future Symposium, Paris, France, 11–13 October 2015. Lindsey Publication Number 11T-001 CROSSING CONDUCTOR TLM.
2. Liu, Y.; Yin, H.; Wu, T. Transmission Line on-line Monitoring System Based on Ethernet and McWiLL. In Proceedings of the International Conference on Logistics Engineering, Management and Computer Science, Shenyang, China, 29–31 July 2015; pp. 680–683.
3. Working Group on Monitoring & Rating of Subcommittee 15.11 on Overhead Lines. Real-Time Overhead Transmission-Line Monitoring for Dynamic Rating. *IEEE Trans. Power Deliv.* **2016**, *31*, 921–927. [CrossRef]
4. Zhirui, L.; Shengsuo, N.; Nan, J. Current Status and Development Trend of AC Transmission Line Parameter Measurement. *Autom. Electr. Power Syst.* **2017**, *41*, 181–191.
5. Hu, Y.; Liu, K. *Inspection and Monitoring Technologies of Transmission Lines with Remote Sensing*; Academic Press: Cambridge, MA, USA, 2017.
6. Li, S.; Li, J. Condition monitoring and diagnosis of power equipment: Review and prospective. *High Volt.* **2017**, *2*, 82–91. [CrossRef]
7. Singh, R.; Choudhury, S.; Gehlot, A. *Intelligent Communication, Control and Devices: Proceedings of the ICICCD 2017*; Springer Nature Singapore Pte Ltd.: Singapore, 2018.
8. Deng, C.-J. Challenges and Prospects of Power Transmission Line Intelligent Monitoring Technology. *Am. Res. J. Comput. Sci. Inf. Technol.* **2019**, *4*, 1–11.
9. Wydra, M.; Kubaczynski, P.; Mazur, K.; Ksiezopolski, B. Time-Aware Monitoring of Overhead Transmission Line Sag and Temperature with LoRa Communication. *Energies* **2019**, *12*, 505. [CrossRef]
10. Xing, Z.; Cui, W.; Liu, R.; Zheng, Z. Design and Application of Transmission Line Intelligent Monitoring System. *E3S Web Conf.* **2020**, *185*, 01063. [CrossRef]
11. Chen, H.; Qian, Z.; Liu, C.; Wu, J.; Li, W.; He, X. Time-Multiplexed Self-Powered Wireless Current Sensor for Power Transmission Lines. *Energies* **2021**, *14*, 1561. [CrossRef]
12. Ahmad, M.R.; Esa, M.R.M.; Cooray, V.; Dutkiewicz, E. Interference from cloud-to-ground and cloud flashes in wireless communication system. *Electr. Power Syst. Res.* **2014**, *113*, 237–246. [CrossRef]
13. Hoole, P.R.P.; Sharip, M.R.M.; Fisher, J.; Pirapaharan, K.; Othman, A.K.H.; Julai, N.; Rufus, S.A.; Sahrani, S.; Hoole, S.R.H. Lightning Protection of Aircraft, Power Systems and Houses Containing IT Network Electronics. *J. Telecommun. Electron. Comput. Eng.* **2017**, *9*, 3–10.
14. Borisov, R.K.; Zhulikov, S.S.; Koshelev, M.A.; Maksimov, B.K.; Mirzabekyan, G.Z.; Turchaninova, Y.S.; Khrenov, S.I. A Computer-Aided Design System for Protecting Substations and Overhead Power Lines from Lightning. *Russ. Electr. Eng.* **2019**, *90*, 86–91. [CrossRef]
15. Lysov, N.; Temnikov, A.; Chernensky, L.; Orlov, A.; Belova, O.; Kivshar, T.; Kovalev, D.; Voevodin, V. Physical Simulation of the Spectrum of Possible Electromagnetic Effects of Upward Streamer Discharges on Model Elements of Transmission Line Monitoring Systems Using Artificial Thunderstorm Cell. *Appl. Sci.* **2021**, *11*, 8723. [CrossRef]
16. Temnikov, A.G. Using of artificial clouds of charged water aerosol for investigations of physics of lightning and lightning protection. In Processing of the 2012 International Conference on Lightning Protection (ICLP), Vienna, Austria, 2–7 September 2012. [CrossRef]
17. Makalsky, L.M.; Orlov, A.V.; Temnikov, A.G. Possible mechanism of lightning strokes to extra-high-voltage power transmission lines. *J. Electrost.* **1996**, *37*, 249–260. [CrossRef]

18. Vasilyak, L.M.; Vereshchagin, I.P.; Glazkov, V.V.; Kononov, I.G.; Orlov, A.V.; Polyakov, D.N.; Sinkevich, O.A.; Sokolova, M.V.; Temnikov, A.G.; Firsov, K.N. Investigation of electric discharges in the vicinity of a charged aerosol cloud and their interaction with a laser-induced spark. *High Temp.* **2003**, *41*, 166–175. [CrossRef]

19. Temnikov, A.G.; Orlov, A.V.; Bolotov, V.N.; Tkach, Y.V. Studies of the parameters of a spark discharge between an artificial charged water-aerosol cloud and the ground. *Tech. Phys.* **2005**, *50*, 868–875. [CrossRef]

20. Temnikov, A.G.; Orlov, A.V.; Chernensky, L.L.; Bolotov, V.N.; Pisarev, V.P. Influence of model hydrometeors on the final stage of a discharge from an artificial charged water aerosol cloud. *Tech. Phys.* **2009**, *35*, 1437–1445. [CrossRef]

21. Temnikov, A.G.; Chernensky, L.L.; Orlov, A.V.; Lysov, N.Y.; Zhuravkova, D.S.; Belova, O.S.; Gerastenok, T.K. Application of Artificial Thunderstorm Cells for the Investigation of Lightning Initiation Problems between a Thundercloud and the Ground. *Therm. Eng.* **2017**, *64*, 994–1006. [CrossRef]

22. Temnikov, A.G. Investigation of peculiarities of discharge formation from the system of artificial charged aerosol clouds of negative polarity. *Electr. Power Syst. Res.* **2014**, *113*, 3–9. [CrossRef]

23. Temnikov, A.G.; Zhuravkova, D.S.; Orlov, A.V.; Lysov, N.Y. Outlook Research of Application of Model Hydrometeor Arrays for Artificial Initiation of Lightning and Thundercloud Charge Reduction. *IOP Conf. Ser. Mater. Sci. Eng.* **2019**, *698*, 044039. [CrossRef]

24. Temnikov, A.G.; Orlov, A.V. Determination of the electric field of a submerged turbulent jet of charged aerosol. *Electr. Technol.* **1996**, *3*, 49–62.

25. Vereshchagin, I.P.; Temnikov, A.G.; Orlov, A.V.; Stepanyanz, V.G. Computation of mean trajectories of charged aerosol particles in turbulent jets. *J. Electrost.* **1997**, *40–41*, 503–508. [CrossRef]

26. Temnikov, A.G. Dynamics of electric field formation inside the artificially charged aerosol cloud and in a space near its boundaries. In Proceedings of the 12th International Conference on Atmospheric Electricity, Versal, France, 9–13 June 2003. ThC3-017-196.

27. Temnikov, A.G.; Chernensky, L.L.; Orlov, A.V.; Kivshar, T.K.; Lysov, N.Y.; Belova, O.S.; Zhuravkova, D.S. Peculiarities of the electric field calculation of the artificial thunderstorm cells. *Int. J. Circuits Syst. Signal Process.* **2018**, *12*, 305–311.

28. Rakov, V.A.; Uman, M.A. *Lightning: Physics and Effects*; Cambridge University Press: Boca Raton, FL, USA, 2003.

29. Syssoev, V.S.; Kostinskiy, A.Y.; Makalskiy, L.M.; Rakov, A.V.; Andreev, M.G.; Bulatov, M.U.; Sukharevsky, D.I.; Naumova, M.U. A Study of Parameters of the Counterpropagating Leader and its Influence on the Lightning Protection of Objects Using Large-Scale Laboratory Modeling. *Radiophys. Quantum Electron.* **2014**, *56*, 839–845. [CrossRef]

30. Kostinskiy, A.Y.; Syssoev, V.S.; Mareev, E.; Rakov, V.; Andreev, M.G.; Bogatov, N.; Makal'Sky, L.M.; Sukharevsky, D.I.; Aleshchenko, A.; Kuznetsov, V.; et al. Electric discharges produced by clouds of charged water droplets in the presence of moving conducting object. *J. Atmos. Sol. Terr. Phys.* **2015**, *135*, 36–41. [CrossRef]

31. Kostinskiy, A.Y.; Syssoev, V.S.; Bogatov, N.A.; Mareev, E.A.; Andreev, M.G.; Makalsky, L.M.; Sukharevsky, D.I.; Rakov, V.A. Infrared images of bidirectional leaders produced by the cloud of charged water droplets. *J. Geophys. Res. Atmos.* **2015**, *120*, 10,728–10,735. [CrossRef]

32. Kostinskiy, A.Y.; Syssoev, V.S.; Bogatov, N.A.; Mareev, E.A.; Andreev, M.G.; Makalsky, L.M.; Sukharevsky, D.I.; Rakov, V.A. Observation of a new class of electric discharges within artificial clouds of charged water droplets and its implication for lightning initiation within thunderclouds. *Geophys. Res. Lett.* **2015**, *42*, 8165–8171. [CrossRef]

33. Kostinskiy, A.Y.; Syssoev, V.S.; Bogatov, N.; Mareev, E.; Andreev, M.G.; Bulatov, M.U.; Makal'Sky, L.M.; Sukharevsky, D.I.; Rakov, V.A. Observations of the connection of positive and negative leaders in meter-scale electric discharges generated by clouds of negatively charged water droplets. *J. Geophys. Res. Atmos.* **2016**, *121*, 9756–9766. [CrossRef]

34. Rakov, V.A.; Mareev, E.A.; Tran, M.D.; Zhu, Y.; Bogatov, N.A.; Kostinskiy, A.Y.; Syssoev, V.S.; Lyu, W. High-Speed Optical Imaging of Lightning and Sparks: Some Recent Results. *IEEJ Trans. Power Energy* **2018**, *138*, 321–326. [CrossRef]

35. Mazur, V. *Principles of Lightning Physics*; IoP Publishing: Bristol, UK; New York, NY, USA, 2016.

36. Sharma, S.R.; Cooray, V.; Fernando, M.; Miranda, F.J. Temporal features of different lightning events revealed from wavelet transform. *J. Atmos. Sol. Terr. Phys.* **2011**, *73*, 507–515. [CrossRef]

37. Esa, M.R.M.; Ahmad, M.R.; Cooray, V. Wavelet analysis of the first electric field pulse of lightning flashes in Sweden. *Atmos. Res.* **2014**, *138*, 253–267. [CrossRef]

38. Temnikov, A.G.; Chernensky, L.L.; Belova, O.S.; Orlov, A.V.; Zimin, A.S. Spectral characteristics of discharges from artificial charged aerosol cloud. In Proceedings of the IEEE Conference Publications: Lightning Protection (ICLP), Proceedings of the 2012 International Conference on Lightning Protection, Shanghai, China, 11–18 October 2014. Article number 6973333.

39. Cooray, V. *Lightning Electromagnetics*; IET Publishing: London, UK, 2012.

40. Cooray, V.; Cooray, G. Electromagnetic fields of accelerating charges: Applications in lightning protection. *Electr. Power Syst. Res.* **2017**, *145*, 234–247. [CrossRef]

41. Cooray, V.; Cooray, G. The Electromagnetic Fields of an Accelerating Charge: Applications in Lightning Return-Stroke Models. *IEEE Trans. Electromagn. Compat.* **2010**, *52*, 944–955. [CrossRef]

42. Cooray, V.; Cooray, G. Electromagnetic radiation field of an electron avalanche. *Atmos. Res.* **2012**, *117*, 18–27. [CrossRef]

43. Shi, F.; Liu, N.; Dwyer, J.R.; Ihaddadene, K.M.A. VHF and UHF Electromagnetic Radiation Produced by Streamers in Lightning. *Geophys. Res. Lett.* **2019**, *46*, 443–451. [CrossRef]

44. Gushchin, M.E.; Korobkov, S.V.; Zudin, I.Y.; Nikolenko, A.S.; Mikryukov, P.A.; Syssoev, V.S.; Sukharevsky, D.I.; Orlov, A.I.; Naumova, M.Y.; Kuznetsov, Y.A.; et al. Nanosecond electromagnetic pulses generated by electric discharges: Observation with clouds of charged water droplets and implications for lightning. *Geophys. Res. Lett.* **2021**, *48*, e2020GL092108. [CrossRef]
45. Rochelle, P. ORBCOMM Announces Commercial Service for Its Final 11 OG2 Satellites. 1 March 2016. Available online: https://www.orbcomm.com/en/company-investors/news/2016/orbcomm-announcescommercial-service-for-its-final-11-og2-satellites (accessed on 30 August 2021).
46. Our Mission Is Space Communication. Available online: http://www.gonets.ru/eng/company/mission/ (accessed on 30 August 2021).

MDPI
St. Alban-Anlage 66
4052 Basel
Switzerland
Tel. +41 61 683 77 34
Fax +41 61 302 89 18
www.mdpi.com

Energies Editorial Office
E-mail: energies@mdpi.com
www.mdpi.com/journal/energies